THE Concise Encyclopedia of Favorite WILD FLOWERS

A HUNDRED NATIVE PLANTS TO APPRECIATE IN THE WILD AND TO GROW IN THE HOME GARDEN

Illustrated by Katharine Burton
Photographs by the Author

MARJORIE J. DIETZ

EDITOR, FLOWER GROWER, THE HOME GARDEN MAGAZINE: AUTHOR, THE CONCISE ENCYCLOPEDIA OF FAVORITE FLOWERING SHRUBS AND THE CONCISE ENCYCLOPEDIA OF FAVORITE FLOWERS

Weathervane Books
New York

Library of Congress Catalog Card Number: 65-10540

This edition is published by Weathervane Books
a division of Barre Publishing Co., Inc.
by arrangement with Doubleday & Co., Inc.
a b c d e f g h
Manufactured in the United States of America

For my father, who, while in pursuit of spring's earliest trout, also was aware of spring's earliest wild flowers, a few of which he usually brought back for my first wild-flower garden.

CONTENTS

Introduction vii

Alum-root 1
Anemone 4
Aster 8

Baneberry 12
Bead-lily 15
Bearberry 18
Beard-tongue 21
Bergamot 25
Bishop's-cap 28
Bitter-root 30
Black-eyed Susan 33
Bloodroot 36
Bluebell 39
Blue Cohosh 41
Blue-eyed-grass 43
Bluets 46
Bugbane 49
Butterfly-weed 52

Catchfly 55
Columbine 58
Corydalis 61
Creeping Snowberry 64
Crinkleroot 66
Culver's-root 68

Dicentra 70

Eupatorium 74

Fairy Bells 78
Fairywand 81
False Lily-of-the-valley 83
False Solomon's-seal 85
Fireweed 92
Foam-flower 94
Fringed Orchid 96

Galax 100
Gayfeather 103
Gentian 106
Ginseng 110
Golden-aster 113
Golden Ragwort 116
Goldenrod 118
Goldenseal 121
Golden-star 123
Goldthread 125
Grass-of-Parnassus 127
Grass-pink Orchid 130

Hepatica 132
Hudsonia 135

Indian Cucumber-root 138
Iris 140
Ironweed 144

Jack-in-the-pulpit 146
Jacob's-ladder 149

Lady-slipper 151
Lily 160

Lobelia 165
Lupine 168

Marsh-marigold 171
May-apple 173
Meadow-beauty 176
Meadow-rue 179
Merrybells 182
Milkwort 185
Monkey-flower 188

Oconee-bells 190

Partridge-berry 193
Partridge-pea 196
Phlox 198
Pipsissewa 203
Pitcher-plant 206
Prickly-pear 210

Rattlesnake-plantain 213
Rue-anemone 216

Saxifrage 218
Shinleaf 221
Shooting-star 224
Showy Orchis 227
Skullcap 229
Skunk-cabbage 232
Solomon's-seal 235
Spring Beauty 238
Starflower 241
Swamp Buttercup 247

Trillium 249
Trout-lily 253
Turtle-head 256
Twin-flower 258
Twin-leaf 260
Twisted-stalk 262

Vein-leaf Hawkweed 265
Vervain 267
Violet 270
Virginia-bluebells 275

Wild Calla 278
Wild Geranium 280
Wild-ginger 282
Wild-petunia 285
Wine-poppy 287
Wintergreen 289
Wood-betony 292

Yellow Star-grass 294

Sources of Wild Flowers 296
Index 299

INTRODUCTION

The Concise Encyclopedia of Favorite Wild Flowers, like the two previous volumes in this series, is intended to be a gardener's helper and reference book. It is for those who wish to grow native plants in a semi-naturalistic or informal setting around their homes. It is not a field guide for accurate identification of native plants in the wild—many excellent ones for this purpose can be found in libraries and bookstores.

Interest in our wild flowers has always existed. Horticulture's hall of fame is filled with dedicated persons who pioneered in discovering them, protecting them and evolving proper methods of culture and propagation. Yet even with these concerns and the setting aside of numerous preserves where native plants are protected by law, every year sees further diminishing of the kinds and numbers of wild flowers. The reasons for this decline are obvious. And few states have actual laws that protect existing plantations of native plants. "Don't Pick" lists are suggestions only and rely on the conscience of the individual as to whether they are heeded. Actually, today, the thoughtless picking or indiscriminate digging of native plants has little effect on their decline compared to the huge destruction wrought by bulldozers, which can destroy thousands of plants in an hour. When it is a question of preserving a bogland or dredging the site for a marina, there is no doubt of the final outcome and the ultimate fate of the bog's assortment of native plants—pink Lady-slippers, native Azaleas, False Lily-of-the-valley, Pitcher-plants, and many others.

We can help, though: first by supporting both state and federal programs of conservation, especially proposals that permanently set aside existing wilderness areas and protect them against pressures for their future exploitation. Secondly, we can introduce at least a few plants into our home gardens. Naturally, the more land one owns, the more one can do both to protect plants that are already growing there and to introduce many more that can be purchased from wild-flower nurseries. But small properties with small gardening areas are much more usual today than extensive estates, and, fortunately, even a small place can support an amazing variety and number of wild flowers.

The purpose of this book is to give gardeners a choice among a hundred kinds of wild flowers suitable for growing in today's home garden. There are, of course, many more. Some may already be known to the reader; many others are listed in catalogs of native-plant specialists (see page 296) and in books devoted to the wild flowers of particular regions. Trailing-arbutus and Bunchberry, two of our more popular native plants, have been included in *The Concise Encyclopedia of Favorite Flowering Shrubs.*

Each plant is presented under its most familiar common name; its other common names, as well as its botanical name, are also given, and all will be found in the Index.

MARJORIE J. DIETZ

East Hampton, N.Y.

THE CONCISE ENCYCLOPEDIA OF FAVORITE WILD FLOWERS

ALUM-ROOT

TYPE Herbaceous perennial
FAMILY *Saxifragaceae* (Sax-if-rag-*gay*-see-ee)
GENUS *Heuchera* (*Hew*-ker-uh)
SPECIES *americana* (am-eh-rik-*kay*-nuh)
ORIGIN Ontario to Connecticut, south to Alabama and Louisiana, west to Minnesota

More familiar than Alum-root to many of us is the Coral-bells (*Heuchera sanguinea*) in its improved forms for borders and rock gardens. It, too, is a North American native and is found in Arizona, New Mexico and Mexico. There are many other species similar to Alum-root and Coral-bells, but Alum-root is perhaps the most available and suitable for introduction to wild-flower gardens.

USES Alum-root is effective in colonies in open woodland or in partly shaded rock gardens, on shaded banks, or among scattered boulders. Individual plants show well against logs, against large tree trunks and in the foreground of informal shrubbery groupings. The plants mix well with Ferns.

HABIT OF GROWTH Each plant forms a heavy rosette of foliage, 8 inches or so high, and produces one or more flowering branches which may grow to 3 feet. The common name refers to the thickened root, which is astringent.

FLOWERS The tiny flowers, barely ¼ of an inch long, are produced in a branching panicle toward the top of the tall stem, and while they are hardly beautiful, they command attention because of their number and the height of the stem over the foliage clump. They are long-lasting and remain effective in their seed stage. The flower color is dull brown or bronze; blooming time is late spring and early summer.

FOLIAGE Alum-root has attractive leaves, rounded, lobed and toothed and very similar to those of its near relatives, Foam-flower and Bishop's-cap. Each leaf is about 2 inches or more in width and has a soft, slightly hairy texture. The fresh green color becomes dark red in the fall.

ASSETS Alum-root is of value for its foliage. It is easy to establish and is tolerant of dry woodlands.

FAULTS The flowering stalks which rise with such purpose from the foliage clumps are disappointing when the flowers finally open. They just really aren't much!

CULTURE Alum-root is easily established in average soil, acid or nearly neutral, in open woods or partly shaded situations. Plants are readily divided in the spring or fall and are easily collected from the wild. Most wild-flower nurseries list plants.

ANEMONE

TYPE Herbaceous perennial
FAMILY *Ranunculaceae* (Ran-nun-kew-*lay*-see-ee)
GENUS *Anemone* (An-*nem*-on-ee)
SPECIES *canadensis* (kan-ad-*den*-siss), *caroliniana* (ka-rol-in-ee-*ay*-nuh), *patens* (*pay*-tenz), *quinquefolia* (kwin-kwef-*foh*-lee-uh), *virginiana* (vir-jin-ee-*ay*-nuh)
ORIGIN See Species, below

The Anemones or Windflowers are many, varying greatly in appearance and appeal. There are species native to Asia, Europe and North America; of the latter, only a handful have been included here. A close relative is the Rue-anemone, *Anemonella thalictroides,* which is discussed separately.

USES The Windflowers are suitable for sunny or lightly shaded rock gardens or walls. They can be planted in drifts or colonies along paths or in open woods or at the base of trees and in the foreground of shrub groupings. Some, especially the Pasqueflower (*Anemone patens*), contrast well with early, spring-flowering bulbs and yellow Primroses which have been naturalized.

HABIT OF GROWTH The stems of varying heights and thicknesses, according to the species, rise from tuberous or swollen roots which creep underground to form colonies.

FLOWERS The flowers, flat or cup-shaped, have petal-like rays or sepals, which are white or pastel-colored. The species listed here bloom in spring, with a few producing flowers into summer. The quality of the flowers varies from delicate to almost fragile, as in *A. quinquefolia,* illustrated, to the jewel-like beauty of *A. caroliniana.* The seed heads of some species are decorative.

FOLIAGE Dissected or divided leaves with short petioles, similar to those of the Buttercup and other plants in the same family.

ASSETS A generally intriguing group which can be readily established in wild-flower gardens. Most are easy to transplant from the wild and are also available as seeds or plants from wild-flower specialists.

FAULTS None really, except it often appears that colonies of Anemone show more foliage than flowers!

CULTURE All of the Windflowers included here should be grown in lightly shaded or sunny situations in soil to which quantities of leafmold or peat moss have been added. The tuberous or swollen roots should be planted just beneath the surface. Colonies can be increased by separating the roots in early spring or after flowering.

SPECIES *Anemone canadensis.* Canada Windflower. Canada Anemone. Round-leaved Anemone. Broad-leaf Anemone. Meadow Anemone. Native over a wide area of North America, from Nova Scotia to Alberta, south to the Carolinas, Arkansas and Kansas. Pure white flowers, about 1½ inches across, appear from spring onward, often into late summer. Its height is from 1 to 2 feet. Tolerant of open, dry woods, where it will form colonies. *A. caroliniana.* Carolina Anemone. Native from Nova Scotia to Florida, westward to Texas and Illinois. Its flowers of many rays may be creamy white or purple-blue, and some in-between rose-colored flowers may be found. Blooming time is from early to mid-spring; height, from 4 to 8 inches. This Anemone will please even the most critical, and as it is tolerant of varying soils so long as the drainage is good, it can be successfully introduced into most wild-flower gardens. It will grow in light shade or sun and is a special gem for rock gardens. *A. patens.* Prairie-smoke. Pasque-flower. This lovely Anemone, similar to the European species, *A. pulsatilla,* which is widely grown in this country in rock gardens and in the foreground of flower gardens, is native over a wide area of our Midwest. It is the state flower of South Dakota. The Crocus-like flowers, about 2 inches across, are purple and appear on short, 3- to 6-inch stems in spring. The stems become taller as the flowers fade and are followed by feathery fruits similar to those produced by many Clematis species. Another choice species, especially effective among clumps of naturalized Daffodils. It will grow in sun or very light shade in average, even dry soil.

A. quinquefolia. American Wood Anemone. Windflower. (Illustrated) Native over North America in woodlands. A pretty species with delicate white or pink-tinged flowers, about 1 inch across, on slender, 4- to 8-inch stems, in early spring. Although it is considered shy in its blooming habit, colonies which have been permitted to spread in acid leafmold soils can be most effective.

A. virginiana. Thimbleweed. Native from Nova Scotia to South Carolina and Kansas. The white or green-tinged flowers are about 1½ inches across and appear in late spring and summer. The seed heads which follow give this species its common name. Height of the flowering stems is 2½ feet. A coarser species which nevertheless is of interest when naturalized in dry, acid soil in open woods.

ASTER

TYPE Herbaceous perennial
FAMILY *Compositae* (Kom-*poz*-it-ee)
GENUS *Aster* (*Ass*-ter)
SPECIES *cordifolius* (kor-dif-*foh*-lee-us), *divaricatus* (div-ar-ik-*kay*-tus), *ericoides* (ehr-ik-*koy*-deez), *kumleini* (kum-*lye*-nye), *laevis* (*lee*-vis), *lateriflorus* (lat-er-if-*floh*-rus), *macrophyllus* (mak-roh-*fill*-us), *novae-angliae* (noh-vee-*an*-glih-ee), *novi-belgi* (noh-vye-*bell*-jye), *ptarmicoides* (tar-mik-*koy*-deez), *umbellatus* (um-bel-*lay*-tus)
ORIGIN See Species, below

If the suggested Aster species above seem excessive in number, just consider that of the more than 200 species known in North America, these are just a handful of the good ones. No doubt exception can be taken to this selection, as some can be weedy. However, they are all available from nurseries, which puts the gardener who wants to know which species he is planting a step ahead. (Identification of Asters is difficult even for botanists.) For those who aren't concerned about the specific identity of their plants, every area abounds in Aster wealth, just waiting for the digging. Fortunately many can be safely dug in full bloom; or plants can be marked when in flower in the fall and lifted in spring, the best season for Aster planting. All one has to do with such collected plants is to note the conditions under which they are naturally thriving and duplicate them in the home wild-flower garden. (An old name for Aster is Starwort, for the shape of the flowers.)

USES There are Asters for sun and dry soil, sun and moist soil, shade and dry soil and kinds suited to rock garden and wall planting as well as informal borders with other native plants such as Goldenrod, Golden-aster, Joe-Pye Weed and other late-summer flowering natives.

HABIT OF GROWTH Bushy or clumpy plants, although some send up only one flowering stalk from a rosette of leaves. Heights vary greatly according to the species and growing conditions.

FLOWERS Daisylike or starry, with white, blue, purple, pink or red-purple rays and bright yellow centers. Most native Asters have single flowers; the doubled flowers are found more generally in garden hybrids. Blooming time is late summer and fall, although there are exceptions. *Aster ptarmicoides,* for instance, begins flowering in late June and July, continuing into fall.

FOLIAGE It varies among the species, some, as in the case of *A. ericoides,* being very narrow (Heathlike), or decidedly heart-shaped (*A. cordifolius*) or large and coarse (*A. macrophyllus*). The leaves are arranged alternately.

ASSETS Asters are one of the delights of autumn. They are versatile and adaptable, generally need no fussing once established. The colors of many are exceptionally pure and brilliant; the flower forms, exquisite.

FAULTS Many Asters are weedy, yet even these species have appeal and often add the only color to waste places where so-called better plants can't grow. The flowers are generally poor cutting material.

CULTURE Asters are adaptable to many soil and exposure situations, many thriving in dry woodlands, others, like the New England and New York Asters, preferring rich, moist soil in full sun. Once introduced, they readily self-sow, and established clumps can easily be divided in the spring.

SPECIES *Aster cordifolius.* Blue Wood Aster. Violet-blue flowers on 1- to 4-foot stems from September to October. Native in woods from Quebec to Minnesota and Kansas, south to Georgia and Missouri.

A. divaricatus. White Wood Aster. White flowers, about ½ inch across, in flat-topped heads on angled stalks about 1 to 2 feet tall, in late summer to early fall. Native from Maine to Ontario, south to Georgia and Alabama. This species likes dry woods.

A. ericoides. Heath Aster. Small white flowers and narrow leaves on 2-foot stalks. Common species in waste areas but valuable for its profusion of flowers and tolerance of dry, lean soil. Native from Maine to British Columbia, south to Georgia, Alabama, Texas and Arizona.

A. kumleini. Blue to purple flowers on yellow stems in September. Height about 15 inches. Native from Wisconsin to Texas.

A. laevis. Smooth Aster. Blue-violet flowers on 1- to 3-foot stalks in late summer, early fall. Native from Maine to British Columbia, south to Georgia, Kansas and New Mexico. Tolerant of dry soil.

A. lateriflorus. Calico Aster. Many small white flowers on 2- to 3-foot stems in late summer. Tolerant of dry soil in light shade or sun. Native from Quebec to Minnesota, south to northern Florida, Tennessee and Texas.

A. macrophyllus. Large-leaved Aster. Violet or white flowers on 3-foot stalks with large, heart-shaped leaves of a rough texture. It flowers in September, likes open woods. Native from Quebec to Minnesota, south to North Carolina and Alabama.

A. novae-angliae. New England Aster. (Illustrated) This and the following species are very decorative and have contributed toward modern hybrid Asters cultivated in gardens. (In England, garden Asters are called Michaelmas-daisies.) This Aster has large flowers, violet to red-purple, 1½ inches across, in late summer. It prefers moderately moist, sunny situations. Natively it may be found in wet meadows, and grows very tall. Found from Quebec to South Carolina and Colorado. Easily transplanted from the wild.

A. novi-belgi. New York Aster. Differs from the preceding in having essentially smooth leaves. Flowers vary in color from purple and violet to rose and white. This species also prefers good, moist soil and full sun. Native from Newfoundland and Quebec to Georgia.

A. ptarmicoides. White Upland Aster. White flowers in flat-topped clusters from early summer through fall. Height about 2 feet. A good, long-blooming species for dry, slightly acid to neutral or alkaline soil in full sun. Native from Quebec and New England to Wyoming, south to Georgia and Alabama.

A. umbellatus. White Aster. This tall Aster (up to 5 feet or more in moist, rich soil) has many small white flowers in August and September. A good companion for other late summer natives in rich, moderately moist soil. Native from Newfoundland to Georgia and Iowa.

BANEBERRY

TYPE Herbaceous perennial
FAMILY *Ranunculaceae* (Ran-nun-kew-*lay*-see-ee)
GENUS *Actaea* (Ak-*tee*-uh)
SPECIES *pachypoda* (pak-ki-*po*-duh), *rubra* (*roob*-ruh)
ORIGIN See Species, below

The Baneberry (its intriguing fruits are poisonous) or Cohosh as it is also called is an old favorite. Both of its species should be in every naturalized planting.

USES Excellent plants to establish in the shady woodland garden, where they form bold clumps. Or set plants along an informal stone wall intermingled with Ferns, Primroses and other shade-tolerant plants.

HABIT OF GROWTH Bushy clumps with a strong, fibrous root system; the flowering and fruiting stalks reach to 2 feet or so.

FLOWERS Small white flowers carried in dense, rounded or oblong racemes, about 2 to 3 inches in length, from spring to early summer. While the flower racemes give an attractive, lacy effect, the Baneberry's value is in the autumn display of the prominent berries, which are most unique. Over a dozen or more of the ½-inch berries, shiny white or red, are carried on short horizontal stems in each raceme. It is easy to see why this plant in its white form is called "Doll's-eyes" in New England, as each berry has a dark spot at its end, completing the resemblance to the eyes put in old-fashioned dolls.

FOLIAGE Compound leaves that are divided and toothed.

ASSETS The spring flowers and, especially, the handsome berries in fall. The plants are easy to establish in shade and are offered by all wild-flower specialists. Plants can be collected from the wild in early spring or fall.

FAULTS Children should be warned that these friendly-appearing berries are to be looked at only and not eaten, as they are poisonous.

CULTURE The Baneberry must be grown in shade, in humus-rich, acid or neutral soil. When planting, incorporate leafmold and/or peat moss in the soil. The clumps are easily divided in early spring or fall. Seeds, which should be removed from the pulp of the berry, can be sown in the fall and should germinate in the spring.

SPECIES *Actaea pachypoda.* White Baneberry or Cohosh. (Illustrated) This is the species that is listed in old books and in most present-day nursery catalogs as *Actaea alba.* It is native from southeastern Canada to Georgia, Louisiana, and Oklahoma, in rich woodlands. Its white berries with purple spots at their tips gleam like china, and are carried on dark stems.

A. rubra. Red Baneberry. Native from Nova Scotia to New Jersey and Pennsylvania, west to Minnesota and Missouri. The berries are a shining crimson red. A similar species, *A. arguta,* has smaller berries and is found from Alaska to California and New Mexico.

BEAD-LILY

TYPE Herbaceous perennial
FAMILY *Liliaceae* (Lil-ee-*ay*-see-ee)
GENUS *Clintonia* (Klin-*toh*-nee-uh)
SPECIES *borealis* (bor-ee-*ay*-liss), *umbellata* (um-bel-*lay*-tuh)
ORIGIN See Species, below

Among the loveliest of all natives are these not-so-easy members of the Lily family. One species, *Clintonia borealis,* is better known by its other common name, Bluebeads. (Westerners know two similar species, *C. andrewsiana,* the Red Clintonia, which is native to California's coastal ranges, and *C. uniflora,* Queen-cup, which is found in woodlands from British Columbia through the Pacific Northwest.)

USES Strictly woodland plants for cool shade, where they make attractive colonies. Good companions include Bunchberry (*Cornus canadensis*), Ferns, Trilliums, Partridge-berry, Wintergreen, many of our native Orchids and the Twin-flower.

HABIT OF GROWTH The Bead-lilies resemble native Orchids in their growth habits. Two or three or more broad leaves fan out from the soil in a rosette from the underground rhizomes. The flowering and fruiting stalk rises above the foliage to 1½ feet or less.

FLOWERS Exquisite, nodding bells in terminal clusters which leave no doubt that they belong to the Lily family. Flowering time is early summer. The fruits which follow can be spectacular, each flower producing one beadlike berry on a short stem.

FOLIAGE Handsome green blades of an elongated oval shape which grow directly from the root.

ASSETS Perfection in foliage, flowers and fruit.

FAULTS Not easy to grow unless conditions can be provided that virtually duplicate their natural haunts.

CULTURE The Bead-lilies naturally grow in cool, deep-shaded woods in rich soil that is always moist. If acid soil of a high humus content—add quantities of leafmold and/or peat moss to a clayey or sandy loam—can be provided as well as cool, full shade, they should thrive in a wild-flower garden. Hot, humid summers are not to their liking. (The author has found that *Clintonia borealis* performs quite well in an artificial bog. The bog was made as an oasis in otherwise dry, acid woodland for native plants which need constant moisture by excavating a rectangular area about 9 feet by 5 feet to a depth of about 2 feet. The excavation was then lined with two layers of medium-gauge polyethylene sheeting and filled with a mixture of peat moss, sand and compost and an inch-layer of sphagnum moss on the top. A tile drain was added to drain off surface water. The edge of the bog was disguised by large, half-rotted Oak logs.) The Bead-lilies are not difficult to transplant from the wild in generous clumps in spring or fall, but keeping them permanently happy may be a problem. Once thriving and established, Bead-lily clumps may be readily divided by separating the rhizomes in early spring or in fall. Seeds may be sown in fall for spring germination.

SPECIES *Clintonia borealis.* Yellow Bead-lily. Bluebeads. Corn-lily. Blue Bead-lily. Whatever its common name, this plant is a special gem, perhaps best appreciated when it is observed in dark woods where it grows like a weed, its leaves springing from rock crevices, old logs and virtually covering the rich earth of swamps. Its range includes Labrador to Minnesota, south into the highlands of Georgia, Indiana and Wisconsin. It produces three to six nodding, green-yellow flowers in clusters atop an 8- to 12-inch stem above the leaves, 4 to 12 inches long. The beadlike bright blue berries which follow the flowers are lovely in mid- to late summer.

C. umbellata. Speckled Bead-lily. Speckled Wood-lily. White Bead-lily. Native from New York to Georgia in rich woodlands. Its small white fragrant flowers, about ⅕ of an inch long and speckled with purple, appear in an umbel above foliage similar to that of the preceding species but covered with fine hairs. Black berries follow the flowers. Although less showy than *C. borealis* and the Western species, it is more tolerant of summer heat. It needs the same rich, acid soil, which should not be allowed to dry out in the summer.

BEARBERRY

TYPE Evergreen sub-shrub
FAMILY *Ericaceae* (Ehr-ik-*ay*-see-ee)
GENUS *Arctostaphylos* (Ark-toh-*staff*-il-os)
SPECIES *uva-ursi* (yew-vuh-*oor*-sye)
ORIGIN North America, Europe, Northern Asia

Bearberry or Kinnikinnick is most appreciated when it forms vast sheets of shining greenery over sand dunes or cliffs and in mountainous situations. Yet it can be tamed and introduced to more restricted areas of many home wild-flower gardens. Bearberry is in the same family as Rhododendron, Azalea, Mountain-laurel, Trailing-arbutus and Heather. The species listed here is one of our most widely distributed native plants. Other species of Bearberry are native to the West Coast states; California in particular possesses a number of kinds. Westerners usually call Bearberry by the name of Manzanita.

USES A superb ground cover, attractive at all seasons. Bearberry is often the one solution for difficult situations, and yards and yards of its handsome trailers will clothe sand dunes, exposed hills and rocky areas where few other plants will grow. Small plants can be set out in dry, sandy pockets in sun-baked rock gardens and kept in bounds by pruning.

HABIT OF GROWTH A trailing or creeping plant, sending shoots far from its main woody stems, which are often hidden by the leaves. Shoots in contact with the soil eventually root.

FLOWERS White or pink, urn-shaped bells like those of the Blueberry, about ¼ inch long, in spring over a long period. They are followed by showy, bright red berries about ¼ inch wide, which are favorites of many kinds of wild life.

FOLIAGE Small, glossy leaves, evergreen and alternate, crowded along the stems.

ASSETS The glistening leaves, always attractive, the cheery red berries in fall and winter, and the usefulness of the plant in soil and situations often too inhospitable for anything else.

FAULTS Plants are difficult to transplant from the wild. Bearberry must have an acid, well-drained soil.

CULTURE Bearberry is naturally found in acid, light, dry soil, even in pure sand, although it can be established in better soils, such as a sandy-peat soil which supports Heaths (Erica) and Heather (Calluna), and dwarf, small-leaved Azalea and Rhododendron plants. A clump may look deceptively easy to lift from the wild, but its fine, hairlike roots, as well as its aboveground runners, travel far and deep. Thus trying to dig clumps in spring or fall results in such mutilation that plants often do not recover. Some success can be achieved by digging hunks or mats of the plants in the winter when the ground is partially frozen. Even though parts of the plants die, new growth appears in the summer. A similar method is to dig up rooted runners in winter and partially bury them in a narrow trench in their new location. Much more foolproof is taking cuttings, 2- to 4-inch shoots in early to midsummer. Insert these cuttings in flats filled with an equal mixture of peat moss and sand, soak thoroughly, then cover the flats with tents of polyethylene and keep in a partly shaded area for four weeks or so. No additional watering will be necessary until the cuttings are set out in early fall or the following spring. Nurserymen put such rooted cuttings into small pots, but there is no reason why the rooted cuttings can't be set out directly where they are to remain in the home garden. They will need frequent watering at first until the roots become established. The cuttings grow slowly but soon send runners forth in all directions. In fact, in a small garden, some pruning back of the longer shoots may be necessary.

BEARD-TONGUE

TYPE Herbaceous perennial
FAMILY *Scrophulariaceae* (Skrof-yew-lay-ree-*ay*-see-ee)
GENUS *Penstemon* (Pen-*steem*-on)
SPECIES *angustifolius* (an-gus-tif-*foh*-lee-us), *cobaea* (ko-*bee*-uh), *digitalis* (dij-it-*tay*-liss), *grandiflorus* (gran-dif-*floh*-rus), *hirsutus* (herss-*yew*-tus)
ORIGIN See Species, below

The lovely Beard-tongues have charmed—and exasperated—gardeners for years. There are so many of them, especially in the prairie and western states, some of which do fairly well when removed from their regions while others may languish. Even the hybrids and cultivated kinds most often planted in manicured flower borders and rock gardens may perform well one year and disappear the following season. Probably some of these problems are due to the particular nature of the species or hybrids, which may be annual or biennial; yet the fact remains that many Beard-tongues just aren't happy beyond their own regions. Those who live within the vast areas where so many of our most beautiful species are native can have a heyday with Beard-tongues. The American Penstemon Society (Secretary, Mrs. Andrew Dowbridge, 25 Auburn Street, Springvale, Maine) is a wonderful source of help for those interested in this genus. Membership costs $2.00 a year. Most of the more familiar species have been omitted here because they are more suitable for cultivated borders than for naturalistic plantings.

USES The sun-loving Beard-tongues are fine in borders, rock gardens and rock walls, and on slopes and banks. The open-woodland types should be colonized at the edge of woods or under high trees where they receive some sun for at least a portion of the day. All Beard-tongues are attractive in flower arrangements.

HABIT OF GROWTH Upright or semi-decumbent clumps, the types considered here, growing 2 to 3 feet in height.

FLOWERS Double-lipped, tubular flowers in racemes or panicles. Some of the large-flowered types approach the Gloxinia in size and color effects, and most gardeners will notice the close resemblance the flowers bear to the Foxglove, a near relative. While perhaps the prettiest Penstemons are in the blue and purple range, there are white, pink, rose and red types. The species here are white or in the blue-purple range. Flowering time is summer.

FOLIAGE Generally quite handsome and differing widely as to texture and color among the species. Usually there is a pronounced rosette of leaves at the base, with pairs or opposite leaves continuing up the flower stems.

ASSETS Many of the most beautiful North American plants are Beard-tongues, and their flowers can be spectacular. Some are quite tolerant of drought and lean soil. They are prolific with their seeds.

FAULTS Many Beard-tongues tend to be short-lived and temperamental outside of their regions.

CULTURE It is impossible to give blanket growing recommendations for all the Beard-tongues. The species listed here generally need full sun or part shade, and a well-drained, sandy soil. All the Penstemons are easily grown from seed, which can be started in early spring or in midsummer after harvesting. In fact, Penstemons self-sow with abandon when conditions please them. Clumps are also readily divided in spring.

SPECIES Those listed below are a mere sample of the riches offered by Penstemon. Consult catalogs of specialists as well as general nursery catalogs and publications of the American Penstemon Society.

Penstemon angustifolius. Native from South Dakota to Montana and New Mexico. Lovely, deep azure flowers on 2-foot stalks in early summer. Its foliage has a bluish cast. Well worth trying in sandy soil beyond its range. Both plants and seeds are available.

P. cobaea. Native from Missouri to Texas. Large purple flowers on 2-foot stems. Several hybrids or selections are available. Needs sun, moderately rich soil with lime.

P. digitalis. (Illustrated) Native over a wide area, from Maine to South Dakota and Texas. This is a very tolerant species and, as luck would have it, is not especially showy. Nevertheless, a colony of plants with their white flowers makes a pleasant accent in summer in open woodland. The plants grow from 3 to 4 feet tall and readily self-sow once established.

P. grandiflorus. Native from Illinois to North Dakota, Wyoming and Texas. Beautiful—but tricky beyond its dry prairie range—

with large, deep-lavender flowers and attractive glaucous foliage. It flowers in early summer. Well worth fussing over, either starting with plants or seeds. Its flowers are a special joy.

P. hirsutus. Native from Maine to Virginia and Wisconsin. Reliable for sun and average, well-drained soil. It has dark purple flowers on 1- to 2-foot stems. Seeds and plants are available.

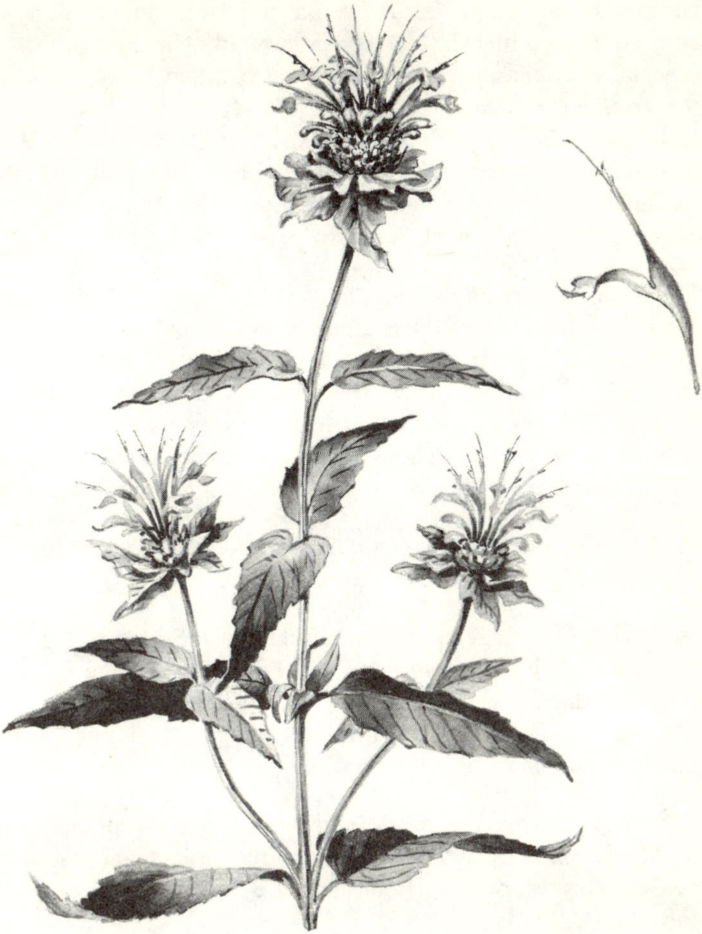

BERGAMOT

TYPE Herbaceous perennial
FAMILY *Labiatae* (Lay-bee-*ay*-tee)
GENUS *Monarda* (Mon-*nard*-uh)
SPECIES *didyma* (*did*-im-uh), *fistulosa* (fiss-tew-*loh*-suh)
ORIGIN Eastern half of North America

Bergamot, also known as Bee-balm and Horse-mint, has always been widely cultivated. The red-flowered *Monarda didyma* is especially familiar. (This species is also called Oswego-tea after the tribe of Indians who made a beverage from the leaves.) Less known is *M. fistulosa,* also called Bergamot or Horse-mint, which makes lovely drifts of lavender along country byways in midsummer. A very similar genus in western North America is Monardella, there called Pennyroyal.

USES The Bergamots are the perfect natives for naturalizing for bold effects. Drift them along split-rail fences, against stone walls, along drives or paths where their Minty fragrance can be enjoyed, in meadows or at the edges of woodlands. They can be attractive along brooks and beside large pools or ponds, either natural or man-made. They make good companions for naturalized Day-lilies. Their flowers are long-lasting in arrangements.

HABIT OF GROWTH Tall, 3 to 6 feet, bushy clumps, with fast-spreading underground stems.

FLOWERS Shaggy flowers in clusters, at the ends of the stalks, in vivid red, lilac or purple. The bracts beneath the heads are tinged with red in the case of *M. didyma,* chartreuse in *M. fistulosa.* There are garden forms of Bergamot in pink, maroon and salmon shades as well as white.

FOLIAGE Aromatic and Mintlike; fragrance is especially strong in *M. didyma.* Leaves are opposite, lanceolate and toothed at the margins; those of *M. didyma* are larger and coarser than those of *M. fistulosa.*

ASSETS The Bergamots are generous with their long-lasting, bright flowers which first appear in early summer and continue for many weeks. They are adaptable to a variety of soils and situations, are easy to establish, either as collected plants or as purchased clumps from a nurseryman. The Bergamots will make a good show the year after they have been planted, and can be readily divided. When spotted in the wild, Bergamots in bloom can be picked freely. The fresh Minty aroma from their foliage reaches even the most stubborn nose.

FAULTS The Bergamots are too rampant to be introduced where they may crowd out more delicate plants. Whatever the temptation, the lovely Lavender Bergamot (*M. fistulosa*) should not be planted in a perennial border where tender loving care and nourishment may send clumps soaring to 6 feet!

CULTURE Growing the Bergamots is about as simple a process as for any native plant. They will tolerate any soil that does not become bone-dry and perform well in part shade or full sun. The best times for planting or dividing are spring and fall, but the collector can lift blooming plants and expect them to survive.

SPECIES *Monarda didyma.* Bee-balm. Bergamot. Oswego-tea. (Illustrated) Native from Maine to Michigan, south to Georgia and Tennessee. It is most often found in moist situations where it may be 5 feet tall, more often between 3 and 4 feet. The bright red flowers are very showy, each individual flower resembling the claw of a cooked lobster. The green leaves are deeply veined, strongly aromatic. Many hybrids are available and should be planted in perennial borders.

M. fistulosa. Lavender Bergamot. Horse-mint. Tall-growing, usually found in semi-shady, moist situations from Maine to Florida and west to Louisiana. Its leaves are smaller than those of *M. didyma,* smoother in texture and far less aromatic. The lavender flowers are also claw-shaped and emit a delicate Mint fragrance when they flower in midsummer. This species is especially effective when naturalized in moist soil and part shade with Day-lilies and the shrub Summersweet (*Clethra alnifolia*) as a background.

BISHOP'S-CAP

TYPE Herbaceous perennial
FAMILY *Saxifragaceae* (Sax-if-rag-*gay*-see-ee)
GENUS *Mitella* (Mit-*tell*-uh)
SPECIES *diphylla* (dye-*fill*-uh)
ORIGIN Quebec to North Carolina and Missouri

Bishop's-cap—also known as Miterwort—is a more delicate version of the better-known Foam-flower. Both of its common names refer to the shape of the seed capsule, which is supposed to resemble the liturgical headdress of a bishop. There are several similar species of Bishop's-cap native to the Far West.

USES Several plants make an attractive ground cover in open areas of a woodland, around trees and as foreground plantings of native shrubs and Rhododendrons. Bishop's-cap will settle happily in a shaded rock garden or in the foreground of shaded sections of a perennial border. It is a good companion to Ferns.

HABIT OF GROWTH Low clumps of foliage from which rise flowering stems to a foot or less.

FLOWERS Delicate racemes of very small, white-petaled, fringed flowers which are arranged sparsely on slender stalks and appear in mid-spring.

FOLIAGE Heart-shaped, lobed and toothed leaves of a hairy texture which form mounds or clumps. Nearly midway on each flowering stalk is found a pair of opposite leaves.

ASSETS A charming plant which is adaptable to a variety of shady situations. Its foliage remains fresh and green all summer.

FAULTS Unassuming, its flowers lacking the splash of many spring-blooming natives.

CULTURE Bishop's-cap prefers shade, which can be quite dense, and a rich, evenly moist soil which may be acid or near neutral. However, it does quite well in dry woods, especially if some water can be provided during drought. Plants can be easily dug from the wild or purchased from nurseries in spring or fall. Clumps are readily divided in spring or fall.

BITTER-ROOT

TYPE Herbaceous perennial
FAMILY *Portulacaceae* (Port-yew-lak-*kay*-see-ee)
GENUS *Lewisia* (Lew-*iss*-ee-uh)
SPECIES *rediviva* (red-iv-*vye*-vuh), *tweedyi* (*tweed*-eye)
ORIGIN Pacific Northwest and Rocky Mountains

A very special native is this lovely Far West plant, first discovered by the Lewis and Clark expedition. The genus, a large one with many more species than those listed here, intrigues rock gardeners. Lewisia is often grown with success far from its native areas, and of course, it is readily cultivated by gardeners of the West who can give it similar growing conditions to those it enjoys in the mountains and surrounding rocky flats. The common name, Bitter-root, really belongs only to one species, *Lewisia rediviva,* the state flower of Montana. Its roots, which were gathered by Indians as food, have a bitter taste that disappears with cooking. Those who can't grow Lewisia can certainly appreciate it when on trips into our western national parks.

USES For rock gardens, slopes and banks, terraces, rock outcroppings and other sites in full sun.

HABIT OF GROWTH Very short-stemmed flowers and narrow leaves grow directly from a high, prominent crown above fleshy roots. Both the crown and roots resemble an octopus.

FLOWERS Beautiful, many-petaled flowers, 2 inches or more wide, almost waxlike in their perfection, and with stems so short that they appear to be sitting on the ground. Flowering time is late spring or in the wake of the receding snow line.

FOLIAGE The leaves of Bitter-root (*L. rediviva*) are narrow and fleshy, appear in spring and usually have shriveled by the time the flowers open. *L. tweedyi* has evergreen leaves.

ASSETS Flowers of exceptional quality and beauty.

FAULTS The Lewisias require a summer baking for their roots, a requirement almost impossible to provide in the East and Midwest.

CULTURE The Lewisias should be grown in what rock-gardening experts call a scree—a site in full sun, its soil, not really soil at all but a deep mixture of stone chips and sand with only a trace of loam or humus. In the case of Lewisia, the stones should not be limestone, as Lewisia is not a lime lover. The next best thing to the scree for Lewisia would be a lean, sandy soil with good drainage. In spring while making leaf and flower growth, Lewisia should have ample moisture; after flowering and as it goes into summer dormancy, the roots should remain as dry as possible. Top growth in the form of foliage may appear again in the fall, when more moisture may be tolerated. Always plant Bitter-root and any other Lewisia species in colonies. Several nurseries offer plants of *Lewisia rediviva.* Seeds of this and other species are also available and should be planted in the fall.

SPECIES *Lewisia rediviva.* Bitter-root. (Illustrated) British Columbia to the Rocky Mountains. Lovely satiny pink flowers about 2 inches across, on 1- to 2-inch stems. The leaves are from 1 to 4 inches long. The specific name, *rediviva,* which means "restored to life," is for the remarkable vitality of Bitter-root's root system which can be dry for months and still spring back into growth when watered.

L. tweedyi. Native to Washington. Spectacularly lovely. Apricot flowers, 2 inches across, on 4-inch stems above fleshy evergreen leaves that are broader than those of *L. rediviva.*

BLACK-EYED SUSAN

TYPE Herbaceous biennial or perennial
FAMILY *Compositae* (Kom-*poz*-it-ee)
GENUS *Rudbeckia* (Rud-*bek*-ee-uh)
SPECIES *hirta* (*hert*-uh)
ORIGIN Widely distributed from Ontario to Florida and Texas

Black-eyed Susan is a native of the Western prairies but has spread so extensively that there is hardly a region where it is not familiar. Recently it has been glamorized by hybridizers, renamed the Gloriosa-daisy and, in this new form, become a popular plant in annual and perennial borders. Still, Black-eyed Susan possesses great appeal in its own right, even though it is considered a weed by farmers. There are several other Coneflowers and Daisies which are often found in the same situations as Black-eyed Susan. Similar species to the Black-eyed Susan include Brown-eyed Susan or Thin-leaved Coneflower (*Rudbeckia triloba*) and Green-headed Coneflower (*R. laciniata*), the single-flowered form of the garden variety "Golden Glow." Wing-stem (*Actinomeris alternifolia*) is often seen in the same field with Black-eyed Susan, but differs from it in having smaller flowers, their rays sweeping down and outward. The leaves clasp the stems to form "wings." There is also Sneezeweed (*Helenium autumnale*), which, in improved forms, is often grown in gardens. While the very familiar Field or White Daisy (*Chrysanthemum leucanthemum*) does not have the prominent cone in the center of its ray flower head as do the preceding, it is common in the same situations, blooming slightly earlier. It is not a true native plant, its original home being Europe, but it has escaped into fields all over this country and is very much at home here.

USES Best allowed to self-sow in sunny fields and byways. Occasional clumps can be introduced into small wild-flower gardens for summer interest. The flowers are excellent for cutting.

HABIT OF GROWTH Black-eyed Susan makes upright clumps with several hairy flowering stems, about 2 feet tall.

FLOWERS The Daisy flowers have a brown, cone-shaped center surrounded by golden-yellow rays, the entire flower head measuring about 2 inches across. The blooming period is a long one, with flowers first appearing in early summer, but some flowers can be found all summer.

FOLIAGE The alternate leaves are coarse and hairy and by late summer may be marred by red-spider infestation.

ASSETS This brazen native is beloved by even the most critical for its bright flowers. It is drought-resistant and tolerates poor soil.

FAULTS Its flowers are its only glory. The rest of the plant is coarse and weedy. It is detested by farmers.

CULTURE Black-eyed Susan prefers dry, poor soil, which is acid. Full sun is best, although it will take light shade. The nature of the plant is to behave as a biennial (growing from seed the first year and flowering the second, then dying after producing its seed harvest), but it often acts as a perennial, persisting for several seasons. At any rate it is free with its seed, and once established, Black-eyed Susan is bound to stay around.

BLOODROOT

TYPE Herbaceous perennial
FAMILY *Papaveraceae* (Pap-av-er-*ray*-see-ee)
GENUS *Sanguinaria* (San-gwin-*nay*-ree-uh)
SPECIES *canadensis* (kan-ad-*den*-siss)
ORIGIN Nova Scotia to Manitoba, south to Florida, Arkansas and Nebraska

The Bloodroot, one of the very earliest of wild flowers to bloom, is so lovely that it is even noticed by otherwise preoccupied trout fishermen. The Bloodroot is especially charming in its stages of growth—when it first pushes through brown leaves with its buds and leaves neatly furled, and when it opens its white flowers which gleam among barely awakening woodland surroundings. The Bloodroot belongs to the Poppy family. The genus contains but one species, perfect enough to satisfy any wild-flower enthusiast.

USES The Bloodroot can be transferred to the home wild-flower garden and established among ground covers such as Ivy, Pachysandra or Myrtle. It is suited to any shaded or partially shaded planting, and colonies can be established in the foreground of such shrubs as Azaleas and Rhododendrons, or along woodland paths. Roots can be planted in rich soil pockets in rock gardens and above and below informal rock walls. In fact, there is hardly a home garden which doesn't provide some space for several plants.

HABIT OF GROWTH The fleshy roots, as thick as or thicker than a pencil, are horizontal, and contain "eyes" like those of Peonies; these produce the short 6-inch stalks terminating in a single flower or, rarely, in two or three. The roots contain a blood-red juice.

FLOWERS Each flower, about 1 to 1½ inches across, is composed of pure white or pale pink petals, usually about eight in number. On cloudy days and toward evening the flowers close, the petals furling upward to assume the upright position they show when they first push through the ground. There is a fully double-flowered form, equally lovely, which is offered by several specialists. The single flowers may last for several days, depending on the weather, while the double flowers are generally longer-lasting. Blooming time is early spring—in April in the vicinity of New York City.

FOLIAGE The gray leaves possess special beauty and provide interest because of their changing character from early spring to midsummer. First they remain curiously furled around the stalk and flowering bud, yet even at this stage show their gray color and prominent network of veins. Gradually the flower and stem push away from the young leaf, which remains at the base of the stalk to unfurl finally as the flower fades to show its rounded, deeply lobed form. The leaf, usually one to a stem, continues to grow until mid- or late summer, eventually to reach 6 inches or more in width. It then begins to deteriorate and has disappeared by fall.

ASSETS A lovely native which flowers early and is long-lasting when spring weather conditions are not severe. It is not hard to grow. Plants can be carefully dug from the wild in early spring or, better yet, purchased from wild-flower nurseries.

CULTURE In its natural haunts, the Bloodroot is found in varying kinds of woods with generally rich soils ranging from neutral to acid in reaction. The thick roots should be planted in spring or fall, in a horizontal position, and covered by 1 inch of soil to which peat moss or leafmold have been added. At least three or more roots should be planted several inches apart to start a colony. The number of "eyes" per root usually determines the number of flowers which will later appear. Bloodroot will self-sow, or seeds can be purchased or collected; seed should be sown in early spring or fall. The roots can be dug in late summer and cut apart, each portion retaining one or more of the "eyes" or buds.

BLUEBELL

TYPE Herbaceous perennial
FAMILY *Campanulaceae* (Kam-pan-yew-*lay*-see-ee)
GENUS *Campanula* (Kam-*pan*-yew-luh)
SPECIES *rotundifolia* (roh-tun-dif-*foh*-lee-uh)
ORIGIN Widespread over North America

This Campanula is known around the world. It is native to parts of Asia and Europe as well as North America. Despite its delicate appearance, it has a rugged, tenacious habit and is found in all sorts of waste places and byways. It is capable of procuring a toehold on narrow rock ledges and in mere cracks of boulders. Other names for it: Bluebells-of-Scotland, Harebell. (The Western Harebell, *Campanula petiolata,* is similar.)

USES Scatter this obliging plant in a variety of situations; it should thrive in all of them. It likes sun or part shade, doesn't mind being blown by the wind or splashed by water. It is perhaps most effective in the company of rocks—in rock gardens, rock walls or on rock-strewn slopes.

HABIT OF GROWTH Threadlike stems from 6 to 24 inches or so, depending on soil and location. These stems possess the strength of wire when it comes to supporting the flowers and surviving in general.

FLOWERS Hanging, bell-shaped flowers, about 1 inch long, in loose racemes, in summer. They are bright blue or purple-blue.

FOLIAGE The basal leaves are decidedly heart-shaped but become Grasslike on the flowering stems.

ASSETS Its pretty flowers, produced all summer, and its adaptability to all sorts of situations. Once introduced, it will reseed itself, yet never becomes a nuisance.

CULTURE Well-drained soil, not too rich, which may be acid or neutral, suits Bluebell. Most nurseries specializing in wild flowers and perennials offer plants, or plants can be collected. Established clumps may be divided in spring or fall. Seeds are available also.

BLUE COHOSH

TYPE Herbaceous perennial
FAMILY *Berberidaceae* (Ber-ber-id-*day*-see-ee)
GENUS *Caulophyllum* (Kaw-loh-*fill*-um)
SPECIES *thalictroides* (thal-lik-*troy*-deez)
ORIGIN New Brunswick and Ontario to Manitoba, south to South Carolina, Tennessee and Missouri

Here is a native plant ideally suited to covering the ground in deep shade. Like the May-apple, which is in the same family, it is not a ground hugger but rather forms a low canopy, about 2 to 3 feet high, over the ground.

USES For colonizing in woodlands and for any informal border- or bank-planting in shade. It is a good companion for Ferns, Baneberry, May-apple, Solomon's-seal and False Solomon's-seal and Goldenseal.

HABIT OF GROWTH Shallow, swollen roots from which the thick stem rises to about 3 feet.

FLOWERS The flowers, in early to mid-spring, are not especially showy, and are produced in a cluster just above the foliage. Each flower, about ½ inch across, is greenish-purple with a bright yellow center. The "blue" of Blue Cohosh comes from the beautiful berrylike seeds which quickly follow the flowers and which are most conspicuous in early summer. Each flower produces two seeds which are about ⅓ inch across.

FOLIAGE The foliage is really a single large leaf which is carried in umbrella fashion halfway up the stalk, but the leaf is so divided as to produce three smaller leaflets which are cut like those of the Meadow-rue.

ASSETS The bold, leafy habit of the plant, its ability to thrive and spread in deep shade, and the lovely blue berry display.

FAULTS Blue Cohosh is not for the small woodland garden, as it may crowd out less robust plants. It is not suited to poor, dry soil.

CULTURE Plant the Blue Cohosh in shade and in humus-rich soil which retains some moisture in periods of drought. Plants can easily be increased by dividing the thick root stocks in early spring or fall. Plants can be collected from the wild in spring or fall or ordered from wild-flower specialists. The seeds can be planted in the fall and should germinate in the spring.

BLUE-EYED-GRASS

TYPE Herbaceous perennial
FAMILY *Iridaceae* (Eye-rid-*day*-see-ee)
GENUS *Sisyrinchium* (Siss-ir-*rink*-ee-um)
SPECIES *angustifolium* (an-gus-tif-*foh*-lee-um)
ORIGIN Widely distributed: Newfoundland west to Saskatchewan and British Columbia; south from Newfoundland to Virginia; also Nebraska, Colorado and Utah

Almost everyone is familiar with this aptly named wild flower which looks like a clump of grass that has been dotted with small blue flowers. The flowers are so pretty and plentiful that they tempt one to pick bouquets, but the little blue flowers promptly curl up into nondescript balls. There are many other species of this genus native to North America, all quite similar, and there's hardly a region where one or more species can't be found. The species considered here is the one listed by most wild-flower specialists and is generally typical of the genus.

USES As a ground cover or accent, in rock gardens and walls, along informal paths, and as a colonizer in meadows or open areas of woodlands.

HABIT OF GROWTH Blue-eyed-grass forms a tuft or clump of long, Irislike foliage which grows from a small, bulblike root.

FLOWERS The blue or purple flowers, about ½ inch across, have yellow centers and look like six-pointed stars. They are arranged in loose clusters, Iris fashion, on a stiff, flattened stalk which rises above the foliage usually 6 to 18 inches or so, depending on soil and location. And also like the Iris, the flowers last but a day, actually only for several hours in the morning. Many more flowers open to take the place of the faded ones. The blooming season, a long one, is from late spring into summer.

FOLIAGE Reedlike or Grasslike leaves in tufts, from 4 to 12 inches or more long, but always lower than the flowering stalks.

ASSETS Easy to establish, useful for its foliage and beloved for its pretty blue flowers which are produced over a long period.

FAULTS The flowers open only on bright mornings and last for a few hours.

CULTURE This native tolerates a wide variation in soil conditions. It will thrive in wet meadows in rich soil, in dry, sandy soil, and it often pops up in waste areas where only the most durable plants survive. It does best in full sun but will perform quite well in open woods and where it receives a few hours of morning sun. Plants are easily increased in early spring or fall by pulling apart the clumps. It is an easy plant to collect and will self-sow.

BLUETS

TYPE Herbaceous perennial
FAMILY *Rubiaceae* (Roo-bee-*ay*-see-ee)
GENUS *Houstonia* (Hoos-*toh*-nee-uh)
SPECIES *caerulea* (see-*rew*-lee-ah), *purpurea* (pur-*pew*-ree-ah), *serpyllifolia* (ser-pil-lif-*foh*-lee-ah)
ORIGIN See Species, below

It's little wonder that Bluets, whose tiny yet numerous flowers form pale blue clouds over the ground in spring, have endeared themselves to generations of nature lovers. While only three species are considered here, there are several other species native to this country with only slight botanical differences from these.

USES As ground cover in sunny meadows or open woods, along pathways or in informal rock gardens. Their effect comes from the profusion of flowers, so they should be established in areas where they can spread at will by self-sowing.

HABIT OF GROWTH Either low-growing, forming tufts (*Houstonia caerulea*), or taller in habit to form clumps 9 to 20 inches high (*H. purpurea*).

FLOWERS Both Bluets and Creeping Bluets (*H. serpyllifolia*) have tiny, four-pointed, star-shaped flowers, pale blue or almost white to sky blue enlivened by a touch of yellow in the corolla. They are produced at the tips of threadlike stems from early to mid-spring. The flowers of *H. purpurea* are tubular and violet or purple.

FOLIAGE Fragile, inconspicuous leaves about ½ inch long in a mat or tuft. The foliage of *H. purpurea,* more prominent, is light green and usually produced in a rosette.

ASSETS The sky-blue flowers of Bluets, twinkling like so many stars strewn over the ground. Even though getting them established in new areas offers a challenge, they are found naturally growing in a number of varying situations, so there should be few home wild-flower gardens where it would not be possible to introduce them. Fortunately they self-sow freely.

FAULTS They are not suitable for dense shade and require an acid soil.

CULTURE Both Bluets and the similar Creeping Bluets prefer open situations or full sun and are most often encountered in meadows where the soil is acid, reasonably moist and rich. On the other hand, these Bluets can surprise one by appearing in rather poor, dry soil and light shade. The Mountain Bluet (*H. purpurea*) makes a more substantial plant and is tolerant of light shade or sun and dry woodland conditions. All species will self-sow or can be divided in early spring or fall. Seeds can be sown in fall or spring.

SPECIES *Houstonia caerulea.* Bluets. Quaker Ladies. (Illustrated) This is a common, widely distributed species, being found over much of the eastern half of North America. Each plant forms a tuft of many hairlike stems carrying the blue flowers, rarely over 5 inches in height.

H. purpurea. Mountain Bluets. Purple Bluets. Native from Maryland to Georgia and Mississippi and Arkansas. A much stockier habit characterizes this species, which is hardy far north of its range. Its height may reach 18 inches.

H. serphyllifolia. Creeping Bluets. This species, native from Pennsylvania to Georgia and Tennessee, forms mats and should self-sow freely.

BUGBANE

TYPE Herbaceous perennial
FAMILY *Ranunculaceae* (Ran-nun-kew-*lay*-see-ee)
GENUS *Cimicifuga* (Sim-iss-*siff*-yew-guh)
SPECIES *americana* (am-eh-rik-*kay*-nuh), *racemosa* (ras-em-*moh*-suh)
ORIGIN See Species, below

The name Bugbane refers to the strong odor of the flowers which repels some insects but attracts carrion flies. Other names include Black Cohosh (for *Cimicifuga racemosa*) and Mountain-rattlebox (for *C. americana* in its seed stage) as well as the rather fanciful name Fairy-candles for both.

USES These tall plants are often included in the shaded rear of cultivated flower borders for late summer display. They make good companions for other tall natives like Ironweed, the tall Asters, Joe-Pye Weed, Fireweed, Lilies, naturalized Day-lilies and some of our tall Ferns. They are suitable for colonizing when space is available at the edge of a woodland, among naturalized shrubs and for occasional accents along open woodland paths.

HABIT OF GROWTH Very tall, stately bushes in the case of *C. racemosa* (up to 6, even 8 feet), but only 2 to 3 feet in the case of *C. americana.*

FLOWERS Long, slender racemes of small white flowers which may be from 6 to 36 inches or more long, produced from early summer to fall, depending on the species.

FOLIAGE The large leaves are divided into leaflets, sharply toothed along their edges.

ASSETS The gently swaying racemes of white flowers are most pleasing in midsummer and show up well against surrounding greenery. The plant's abundant foliage is always attractive.

FAULTS These are large, bushy plants, out of scale for some areas where space is limited. They do best in part shade and will not perform well in dry, poor soil.

CULTURE Grow both species in rich soil deeply prepared, to which generous quantities of compost, leafmold and/or peat moss have been added. The soil should have an acid reaction. The clumps are composed of crowns which can be separated after the plants have been established in one area for a few years. Plants are available from all wild-flower nurseries.

SPECIES *Cimicifuga americana.* American Bugbane. Appalachian Fairy-candles. Mountain-rattlebox. Native from southern Pennsylvania to Georgia and Tennessee. Fragrant, fluffy white flowers in racemes from 2 to 4 feet or more, in late summer and into fall.

C. racemosa. Black Snakeroot. Black Cohosh. Cohosh Bugbane. Native from Maine and Ontario to Georgia and west to Missouri and Wisconsin. A soaring perennial which can be stunning in the right spot. Its long white-flowered racemes may reach 8 feet. Flowering time is early summer.

BUTTERFLY-WEED

TYPE Herbaceous perennial
FAMILY *Asclepiadaceae* (Ass-kleep-ee-ah-*day*-see-ee)
GENUS *Asclepias* (Ass-*kleep*-ee-ass)
SPECIES *tuberosa* (too-ber-*roh*-suh)
ORIGIN Maine to Florida and Arizona

Although the Butterfly-weed is a Milkweed, it far outclasses its subtle-colored relatives when its flowers blaze in summer fields. A few other Milkweeds with intricate but less spectacular flower heads are sometimes worthy of introducing in special situations. They include the Swamp Milkweed (*Asclepias incarnata*), the Purple Milkweed (*A. purpurascens*) and the Blunt-leaved Milkweed, which has interesting wavy leaves. The Butterfly-weed in bloom is a favorite stopping-off place for butterflies, and this combination makes an appealing subject for photographers. Another name for it is Pleurisy-root.

USES Butterfly-weed is often planted in perennial borders. It is attractive in a variety of informal settings: naturalized along sunny paths or drives, as a foreground planting to native shrubs, and as important accents among other summer-flowering native plants combined in naturalistic groups. The flowers last well in arrangements.

HABIT OF GROWTH Bushy clumps with several flowering stalks which may be a foot or less tall, or may grow to 3 feet or more, rising from thick fleshy or woody tubers.

FLOWERS Umbels of bright flowers (usually deep orange, but apricot and yellow forms are prevalent too), appearing from early July into August, with some secondary bloom appearing in fall. Whole fields may be colonized by the Butterfly-weed, which self-sows freely, and great color variations may be found.

FOLIAGE Long, narrow leaves of a leathery texture are arranged alternately along the entire stem.

ASSETS Its beautiful, long-lasting flowers of midsummer, its tolerance for a variety of soils, including very dry, lean ones and its adaptability to many kinds of gardens.

FAULTS Plants in flower cannot be lifted successfully. (See Culture, below.)

CULTURE Most nurseries specializing in perennials and wild flowers list plants and/or seeds of the Butterfly-weed. Seeds may be collected in the fall and sown late to germinate in the spring, or they can be sown in the spring. The seedlings should be transplanted by midsummer to their permanent positions. Older, established plants are not so readily transplanted because of their long, tuberous roots which are brittle and easily broken. However, old clumps can be successfully lifted if one is patient and waits until fall or early spring for the operation. Replant in well-drained soil. Although the Butterfly-weed is often found in sandy, acid soils, it will thrive in neutral and slightly alkaline soils also. (It is the Butterfly-weed that is often spotted in full bloom in the inhospitable soil along railroad tracks.)

CATCHFLY

TYPE Herbaceous perennial
FAMILY *Caryophyllaceae* (Karry-off-il-*lay*-see-ee)
GENUS *Silene* (Sye-*leen*-ee)
SPECIES *caroliniana* (ka-rol-in-ee-*ay*-nuh), *virginica* (vir-*jin*-ik-uh)
ORIGIN See Species, below

Many of the Catchfly clan are so bright and tidy that they look like escapees from a well-tended flower garden rather than true wildings. In fact the two species listed here resemble Sweet William and other Dianthus varieties, which is logical enough, as they belong to the same family. The name Catchfly refers to the sticky coating on leaves, stems and flower buds. There are many other Catchfly species native to various parts of this country. Several beautiful species are native to the Southwest as well as the Pacific Northwest.

USES Plant individual clumps in a lightly shaded rock garden; establish colonies along paths in open dry woodland; set plants at top of retaining walls. The Pink Catchfly (*Silene caroliniana*) flowers at the same time as the Crested Iris and makes an appropriate foreground planting for Azaleas and Rhododendrons.

HABIT OF GROWTH The Pink Catchfly makes a neat tuft, about 5 to 10 inches high, depending on soil and location, its flowers virtually covering the plant. The Red Catchfly or Fire Pink (*S. virginica*) is sprawling in form, its reclining stems from 8 to 12 inches long.

FLOWERS Clusters of five-petaled flowers, 1 to 1½ inches across, each flower from a narrow, tubular "neck," the calyx.

FOLIAGE Long, narrow leaves, opposite, sticky to touch.

ASSETS Their showy, well-formed flowers in midspring and summer, and non-invasive growth habits. They are reasonably easy to transplant from the wild and are carried by most wildflower nurseries. The Pink Catchfly, especially, is tolerant of sandy, rocky, dry soil.

CULTURE Plant in spring or fall; with usual care, plants in flower can be dug. The Pink Catchfly natively grows in acid soil and should therefore be planted in soil enriched with leafmold or peat moss. It prefers light shade at least part of the day. When digging plants, go deep to include the long fleshy taproots. The Fire Pink, or Red Catchfly, grows in neutral, slightly acid or very acid soil, also in light shade, but tolerates full sun if its roots remain cool. Both of these Silenes need good drainage.

SPECIES *Silene caroliniana.* (It has been listed as *S. pensylvanica*). Pink Catchfly. Wild Pink. (Illustrated) It is found in sandy or rocky soil throughout the Northeast, from Maine to Georgia, west to Kentucky. In the New York area it flowers in mid-May. Height, up to 10 inches; flowers, dusty pink. This is a choice native plant, well worth cultivating.

S. virginica. Red Catchfly. Fire Pink. It is found throughout the Northeast and west to Arkansas. Flowering time is late spring and intermittently into summer. Its blazing red flowers are 1½ inches across.

COLUMBINE

TYPE Herbaceous perennial
FAMILY *Ranunculaceae* (Ran-nun-kew-*lay*-see-ee)
GENUS *Aquilegia* (Ak-wil-*leej*-ee-uh)
SPECIES *caerulea* (see-*rew*-lee-uh), *canadensis* (kan-ad-*den*-siss), *chrysantha* (kriss-*anth*-uh), *jonesi* (*jones*-eye), *longissima* (lon-*jiss*-im-uh), *skinneri* (*skinner*-eye)
ORIGIN See Species, below

Just about every part of this world knows at least one form of the Columbine, which has long been a popular subject for rock gardens and cultivated flower borders. North America has been enriched by many species, several of which have contributed their blood to the garden hybrids. A few of the native species are considered here; there are many others that are worth while too.

USES Grow Columbines in sun and light shade, where they will naturalize and probably reseed themselves without any help from the gardener. The taller kinds make good companions for Lilies. All are especially at home in rock gardens, against walls or on slopes or banks.

HABIT OF GROWTH Leafy mounds, which can be several inches or more high or mere tufts in the case of some of the alpine species. The flowers are often carried on tall, branching stems up to 4 feet or so, or the flowering stems may rise just above the foliage mound. Each plant has a definite crown.

FLOWERS The Columbine flower is composed of five petal-like sepals, with spurs (really petals) which can be long or short. Each flower is further accented by the numerous long stamens. The flowers are produced at the ends of the branches, and in the taller kinds they hang gracefully. Flowering time and duration differ according to the species, but blooms generally appear in spring and early summer.

FOLIAGE The attractive compound leaves are divided into segments which can be lobed.

ASSETS Their free-blooming habit of long duration, their tolerance of varying soil conditions, their hardiness, the graceful habit of the plant.

FAULTS The Columbine can be infested by leaf miners, tiny worms which bore between the leaf layers, leaving unsightly zigzag trails. Some Columbines tend to exhaust themselves by blooming. Such plants may make up for this defect by self-sowing.

CULTURE Columbines generally thrive in well-drained, moderately acid soils of average fertility. Sun or light shade suits them. Old clumps can be divided in early spring or fall, but it's really easier to let them self-sow or to sow seed yourself. While still small, the seedlings can be readily transplanted to wherever the gardener wishes. Controlling the leaf miner is a nuisance and probably not worth the trouble in extensive naturalized plantings. Spray with lindane, following the manufacturer's directions, as soon as the first tunnel is seen.

SPECIES *Aquilegia caerulea.* Colorado Columbine. (Illustrated) Native to the Rocky Mountains, this is the state flower of Colorado. Blue-purple and white flowers with 2-inch-long, knobbed spurs appear in late spring. An adaptable, enduring species and an important parent of many garden hybrids. Its height is from 8 to 24 inches.

A. canadensis. Known by a variety of common names, including Wild Columbine, American Columbine and even Eastern Columbine. The last is puzzling because this species is widely distributed, starting in Newfoundland and ending up in Texas. Red and yellow flowers with red, knobbed spurs on 2- to 3-foot stalks in mid- to late spring. One of the best-loved, most graceful species, its 1½-inch flowers nod and bob gently in the slightest breeze. It, too, is adaptable and tolerant of varying conditions, but is generally found in light shade in both moist and dry soil. It self-sows with ease.

A. chrysantha. Golden-spurred Columbine. Native from the Rocky Mountains to Texas. Many long-spurred yellow flowers on a bushy 4-foot plant, from late spring to summer.

A. jonesi. Low-growing alpine from Montana and Wyoming. Its large blue flowers sit on tufted 3-inch plants. For cool but sunny rock gardens.

A. longissima. Native from Texas to Mexico. Pale yellow, long-spurred flowers on 3-foot plants. This sun-loving, drought-resistant perennial will bloom and bloom.

A. skinneri. Native to New Mexico and southward. Greenish-yellow flowers with pale red spurs on 2-foot plants. Summer-flowering.

CORYDALIS

TYPE Herbaceous annual and perennial
FAMILY *Fumariaceae* (Few-may-ree-*ay*-see-ee)
GENUS *Corydalis* (Koh-*rid*-al-iss)
SPECIES *aurea* (*aw*-ree-uh), *sempervirens* (sem-per-*vye*-renz)
ORIGIN See Species, below

There are several other species of Corydalis native to North America, all quite similar to each other and showing their close kinship to Dicentra, which they resemble in both flower and leaf. One of the European species, *Corydalis cava* (*bulbosa*), is being recommended by some wild-flower nurseries for naturalizing in open woodlands. It has rosy-purple flowers on 6-inch stems and flowers in early spring. Corydalis is also known as Fumitory or Fume-root because of the strong nitrogenous odor emitted by the crushed roots.

USES For a variety of situations—open woodlands, rock gardens, retaining walls, slopes and banks and along paths. The plants are especially attractive growing from soil pockets in a retaining wall where their seeds had previously lodged.

HABIT OF GROWTH Upright or slightly sprawling plants with branching flowering stalks from 6 to 12 inches high.

FLOWERS Slender, one-spurred flowers which hang in short racemes at the ends of the branches, usually appearing first in mid-spring with some flowers being produced all summer.

FOLIAGE Gray-green leaves, divided into segments and somewhat resembling those of the Columbine and its relative, Dicentra. The Pink Corydalis (*C. sempervirens*) is a biennial; during its first growing year it forms attractive rosettes of gray foliage.

ASSETS These two Corydalis species are obliging and carefree, cheerfully producing their little flowers all summer. They have pleasing foliage, thrive under various conditions and are free with their seeds.

FAULTS Pretty plants, exquisite at close hand, but bound to be overlooked by those who see nothing but the spectacular.

CULTURE Golden Corydalis (*C. aurea*) thrives in light shade such as is offered by an open woodland or in full sun. It seems to do well in both sandy or humusy soil which is neutral in reaction. It is considered an annual, but apparently many plants live longer than one season. Pink Corydalis is definitely a biennial—that is, its seeds germinate and form a rosette of foliage the first year and produce their flowers the second year. It, too, tolerates sun or light shade and is usually found in moderately acid soil. Seeds of both species can be scattered where they are to flower in spring or fall. A sprinkling of ground limestone can be added to the soil before sowing seeds of the Golden Corydalis. Once introduced, both species will self-sow.

SPECIES *Corydalis aurea.* Golden Corydalis. Native from Quebec to Alaska, south to West Virginia, west to Missouri and California. Yellow flowers, ½ inch long, all summer, in racemes at the ends of branches. Annual or biennial in habit.

C. sempervirens. Pink Corydalis. Pale Corydalis. Native from Newfoundland to Alaska and British Columbia, south to New York and Minnesota. Pink, yellow-tipped flowers about ½ inch long, several in racemes at the ends of the branches. A most useful species for rocky, lightly shaded woodland. It will flower from late spring until fall. Seeds of a white-flowered form are offered.

CREEPING SNOWBERRY

TYPE Evergreen perennial
FAMILY *Ericaceae* (Ehr-ik-*ay*-see-ee)
GENUS *Chiogenes* (Kye-*oj*-en-eez)
SPECIES *hispidula* (hiss-*pi*-dew-la)
ORIGIN Canada, south to North Carolina, Minnesota and Idaho

This little plant is a close relative of Wintergreen (*Gaultheria procumbens*), another creeper, and is most often found in the cool, acid woods of the North under evergreen trees. Both plants belong to the Heath family, which includes Trailing-arbutus, Rhododendron and Mountain-laurel.

USES As a ground cover in shaded woods or in rich soil pockets of a shaded rock garden or wall. It makes a good underplanting to Rhododendrons, Azaleas and Highbush Blueberry. Plants do well around the edges of a shaded pool.

HABIT OF GROWTH A ground-hugging plant, only a few inches high, with trailing or creeping stems which can form fairly tight mats.

FLOWERS Small, bell-shaped flowers, white, about ⅙ inch long, in spring. They are followed by shiny white berries (which give this plant its common name) in late summer. The berry has the same Peppermint flavor as Wintergreen.

FOLIAGE Small, olive-green leaves, alternate, rounded, about ⅓ of an inch long. They are evergreen.

ASSETS Desirable for its evergreen habit and snow-white berries.

FAULTS Not suitable for hot, dry woodlands. It must have an acid soil.

CULTURE The Snowberry is not hard to establish in reasonably cool spots where the soil is acid and rich in humus. The soil should retain some moisture. When planting, mix a good quantity of leafmold and/or peat moss and some clay, if the soil is very sandy, into the planting hole. Maintain a half-rotted leafmold mulch around the plants to keep their roots cool in summer. A light winter covering of leaves and Pine needles also helps. Plants can be increased by removing and setting out rooted runners.

CRINKLEROOT

TYPE Herbaceous perennial
FAMILY *Cruciferae* (Krew-*siff*-er-ee)
GENUS *Dentaria* (Den-*tay*-ree-uh)
SPECIES *diphylla* (dye-*fill*-uh), *laciniata* (las-in-ee-*ay*-tuh)
ORIGIN See Species, below

Crinkleroot is also known as Toothwort; both names refer to the roots, which have toothlike protuberances.

USES For colonizing in the woodland or shaded, informal rock garden among other spring-blooming natives like Dutchman's-breeches and Blue Phlox.

HABIT OF GROWTH Low-growing clumps or mats which spread by underground rootstocks. The edible roots are peppery, rather like the leaves of Water-cress in flavor.

FLOWERS Pretty, rather than spectacular, white or pink-tinged flowers, ¾ inch across with four petals, and several to a cluster. Flowering time is early to mid-spring.

FOLIAGE Rather attractive lobed or notched leaves arranged on the stems in whorls.

ASSETS The dainty flowers can be quite effective in the woodland after the clumps become established.

FAULTS Not for dry woodlands.

CULTURE The Toothworts need shade and moist, rich soil which has been liberally enriched with leafmold and/or peat moss. The soil should have an acid reaction and retain its moisture during dry periods. Both species can be readily lifted and divided in early spring or fall.

SPECIES *Dentaria diphylla.* Two-leaved Toothwort. Crinkleroot. (Illustrated) Native from Nova Scotia and New Brunswick south to the Carolinas, westward to Kentucky and Minnesota. Flowers, white or pinkish, clustered on an 8- to 12-inch stem. The leaves, in pairs, are composed of three leaflets, deeply toothed.
D. laciniata. Three-leaved Crinkleroot. Pepper-root. Native from Quebec to Florida, west to Minnesota, Kansas and Louisiana. White flowers in clusters on 6- to 14-inch stems. There are three leaves in whorls from the stem, each leaf divided into three narrow leaflets which are toothed.

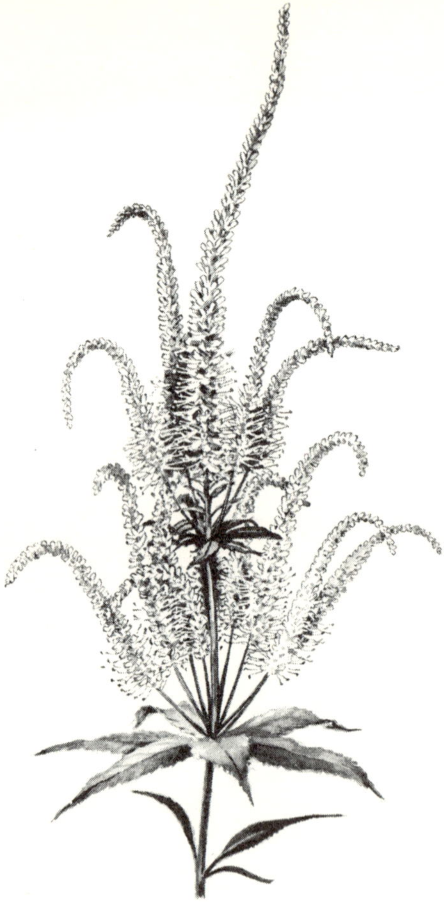

CULVER'S-ROOT

TYPE Herbaceous perennial
FAMILY *Scrophulariaceae* (Skrof-yew-lay-ree-*ay*-see-ee)
GENUS *Veronicastrum* (Ver-on-ik-*kast*-rum)
SPECIES *virginicum* (vir-*jin*-ik-um)
ORIGIN Manitoba, Vermont and Massachusetts, south to Florida and Texas

This widely distributed, good-humored native used to be included among the Veronica species and is still listed as such by some. Its flowers certainly resemble those of many garden Veronicas and its generic name, *Veronicastrum,* means Veronica-like.

USES Another soaring perennial in the class of such giants as Ironweed, Bugbane, and some of the Eupatoriums, especially Joe-Pye Weed. It is suitable with these or by itself wherever a tall accent is required: in open woods, at the edge of woodland, in meadows and along fences.

HABIT OF GROWTH A tall leafy plant, from 3 to 6 feet, which quickly forms large clumps.

FLOWERS White or pink-tinged flowers, small and fuzzy, arranged to form dense spikes, several to a stalk, in summer.

FOLIAGE Sharp-pointed, toothed leaves in whorls up the stems, each leaf about 6 inches long.

ASSETS Culver's-root is a stately, handsome plant which is easy to establish. It has an especially long flowering season, beginning in early summer and extending into the fall.

FAULTS Not for the small, space-restricted wild garden.

CULTURE Culver's-root prefers sun or light shade and moderately moist soil which is acid. Clumps can easily be increased by lifting in spring or fall and dividing.

DICENTRA

TYPE Herbaceous perennial
FAMILY *Fumariaceae* (Few-may-ree-*ay*-see-ee)
GENUS *Dicentra* (Dye-*sent*-ruh)
SPECIES *canadensis* (kan-ad-*den*-siss), *cucullaria* (kuk-yew-*lay*-ree-uh), *eximia* (ex-*im*-ee-uh), *formosa* (for-*moh*-suh)
ORIGIN See Species, below

These are North American relatives of that old-fashioned garden favorite, Bleeding-heart (*Dicentra spectabilis*), which was introduced to the Western world many years ago from Japan. One of our native species, *D. eximia,* is also called Bleeding-heart, its full common name being Fringed Bleeding-heart.

USES Ideal subjects for woodland gardens, under and around trees and shrubs, in rock gardens and informal rock walls, on banks and along paths. All of the species mentioned here, including *D. spectabilis,* are effective in each other's company or in colonies of their own. The Fringed Bleeding-heart makes a wonderful Fernlike ground cover in light shade as it is virtually everblooming.

HABIT OF GROWTH Leafy clumps or mounds from bulbous or tuberous roots, the flowering stalks about 6 to 18 inches tall, depending on the species.

FLOWERS The flowers of all Dicentra species have been favorites of many generations of plant lovers. They are roughly heart-shaped, more markedly so in *D. eximia* and *formosa,* while somewhat resembling a Dutchman's full pants, upside-down, in the case of *D. cucullaria* and *canadensis.* The fragrant flowers wave in loose racemes above the foliage. Blooming time differs according to the species.

FOLIAGE The compound foliage is divided into threes, and can be grayish and almost as delicate as that of many Ferns.

ASSETS Their quaint flowers over lacy foliage clumps. The Fringed Bleeding-heart produces some flowers all summer and is tolerant of varying conditions, including sun or part shade, dry and moist soil. All of these species are available from nurseries.

FAULTS Both Dutchman's-breeches (*D. cucullaria*) and Squirrel-corn (*D. canadensis*) lose their aboveground growth soon after flowering, of no matter except to the careless gardener who may dig into their underground corms.

CULTURE All the Dicentra species appreciate good soil that is moist but well drained, and rich in leafmold. The question of whether it should be neutral or acid seems debatable. Both Dutchman's-breeches and Squirrel-corn are notorious for their shy blooming habit when transplanted from the wild. In very acid woodland soil, the plants may flower better when a light application of ground limestone is made. All of the Dicentras, especially *D. eximia,* self-sow freely, and seedlings may turn up in the oddest places. Established clumps can also be divided; this is easily accomplished by lifting the clumps in early spring or just after flowering and gently pulling apart the tubers.

SPECIES *Dicentra canadensis.* Squirrel-corn. Native from Quebec to Minnesota, south to North Carolina and west to Missouri. The common name refers to the little underground tubers which resemble kernels of corn. Its fragrant flowers, shorter-spurred than those of Dutchman's-breeches, are greenish-white, about ½ inch long, and are carried on 4- to 6-inch arching stems above pretty gray-green leaves in early to mid-spring. Squirrel-corn needs light shade, rich moist soil that is nearly neutral. Add ground limestone lightly to the soil before planting if the soil tests very acid. Fall planting is better than early spring planting for this species.

D. cucullaria. Dutchman's-breeches. White-hearts. (Illustrated) Native from Nova Scotia to North Carolina, west to Missouri, Minnesota and Kansas. As many as ten of its quaint, appealing flowers may dangle from gently swaying stems which may reach a foot in height. Each flower, fragrant and from ½ to ⅔ of an inch long, is white or pink-tinged with a yellow tip. Flowering time is early spring. Its root system is a cluster of little white tubers. Dutchman's-breeches needs conditions similar to those for the preceding species, but it may be more tolerant of moderately acid soils.

D. eximia. Fringed Bleeding-heart. Plumy Bleeding-heart. Wild Bleeding-heart. Western New York to Georgia. This and the following species are part of the parentage of some of the new Bleeding-heart hybrids like 'Bountiful.' Unlike both Squirrel-corn and Dutchman's-breeches, it does not die down after its spring flowering but continues to produce flowers until frost. In fact it is one of our few everblooming native plants. The rosy-pink, short-spurred flowers are produced in racemes, 2 feet or less above the dense mounds of gray, Fernlike foliage. There is a white-flowered form. Wild Bleeding-heart is tolerant of widely varying soils and exposures. It will grow in rich, moist woodland soil that is moderately acid but is also found in dry, sandy woods. It is also tolerant of open or sunny conditions. It self-sows freely.

D. formosa. Western Bleeding-heart. British Columbia to California. Although at its best in meadows of the Sierras, this species will perform fairly well beyond its natural environment. It needs the same conditions as Squirrel-corn and Dutchman's-breeches. It has rose-colored (occasionally white) flowers on stems 12 inches or longer.

EUPATORIUM

TYPE Herbaceous perennial
FAMILY *Compositae* (Kom-*poz*-it-ee)
GENUS *Eupatorium* (Yew-pat-*toh*-ree-um)
SPECIES *coelestinum* (see-less-*tye*-num), *dubium* (*dew*-bee-um), *hyssopifolium* (hiss-op-ee-*foh*-lee-um), *maculatum* (mak-yew-*lay*-tum), *perfoliatum* (per-foh-lee-*ay*-tum), *purpureum* (pur-*pew*-ree-um)
ORIGIN See Species, below

Perhaps the best-known member of this large genus is the stalwart Joe-Pye Weed (*Eupatorium purpureum*), illustrated above, which helps color wet meadows winy-rose in late summer. Also familiar is the white-flowered Boneset (*E. perfoliatum*), found in the same moist fields as Joe-Pye Weed, and the Hyssop Thoroughwort (*E. hyssopifolium*), which forms a haze of white over dry, sandy byways when it flowers in midsummer.

USES All the Eupatoriums are coarse, spreading perennials for open, moist or dry situations (depending on the species), most appreciated when they can form natural colonies. Exceptions are Joe-Pye Weed and Boneset, which will thrive in light shade and can be set out as tall accents in the foreground of a naturalistic shrub planting or at the edge of a woodland. Mist-flower (*E. coelestinum*), often called Perennial-ageratum because of its resemblance to the true, annual Ageratum, can be grown in perennial gardens where it makes a good companion to early-flowering Chrysanthemums. The Hyssop Thoroughwort is especially prevalent in the sunny, sandy fields along the seacoast and should be encouraged as it makes a ground cover where other plants would fail. All of the Eupatoriums are excellent for summer cut-flower arrangements.

HABIT OF GROWTH Tall clumps, often up to 6 feet in the case of Joe-Pye Weed and Boneset, each sending up several flowering stalks.

FLOWERS Flat or slightly rounded terminal clusters made up of small, tubular flowers, from ¼ to ½ inch long, hairy or wispy. Colors include white, purple-pink and blue-purple.

FOLIAGE Generally the opposite leaves are large and toothed, rather coarse, except for those of the Hyssop Thoroughwort, which are narrow and arranged in graceful whorls up the stem.

ASSETS Many of the Eupatoriums are worth introducing (or, if already in existence, encouraging) for the various uses listed above. The dusty-rose or light-purple flowers of Joe-Pye Weed are especially effective with other late summer natives such as

Goldenrod, Golden-aster and Ironweed, and late-flowering naturalized plants of Day-lilies.

FAULTS They are too coarse for most small wild-flower gardens. The Mist-flower, when cultivated in flower borders, needs even, constant moisture.

CULTURE Buy plants from wild-flower nurseries in the spring or dig plants, previously identified and marked, in early spring. Mist-flower, Boneset, and Joe-Pye Weed need rich, moist soil, slightly acid or acid, and a lightly shaded or sunny situation. The Thoroughworts should be planted in spring, in dry, sandy soil in full sun. The Mist-flower spreads readily by stolons and is easily divided in the spring.

SPECIES *Eupatorium coelestinum.* Mist-flower. Perennial-ageratum. Native in open, moist woodlands from New Jersey to Florida and Texas. Its pretty, purple-blue flowers in fuzzy heads are carried atop 2- to 3-foot leafy stalks. The Mist-flower is worth growing for flower arrangements alone. Its spreading habit must be controlled in the flower border or small wild-flower garden.
E. dubium. This is one of several species known as Joe-Pye Weed. It inhabits moist-soil areas from Nova Scotia to South Carolina. Its stems are speckled with purple, may tower to 8 feet. Its broad leaves are carried in whorls of three and four, are coarsely toothed. The dusty-rose or purple flowers appear in flat or slightly rounded heads in late summer.
E. hyssopifolium. Hyssop Thoroughwort. Native to sunny, poor-soil areas from Massachusetts to Florida and Texas. Its white flowers are carried in flat-topped panicles on 3-foot stalks in late summer. Its foliage pattern is interesting and does resemble that of the herb Hyssop, after which it is named. The narrow leaves, from ½ inch to 2 inches long, are bunched in whorls up the stem.
E. maculatum. Spotted Joe-Pye Weed. Its stems are also speckled in purple. It is found in moist soil, often along streams, from Newfoundland to British Columbia, south to Illinois and New Mexico.
E. perfoliatum. Boneset. Native from Nova Scotia to Florida and Texas. It is often found with Joe-Pye Weed and makes a good

companion for the late-summer flowering natives. The white flowers are carried in flat-topped clusters at the end of 4-foot stalks and first appear in midsummer. The opposite leaves are carried in pairs and clasp the stem so that it appears to be passing through them.

E. purpureum. Joe-Pye Weed. (Illustrated) This species is widely distributed from Maine to Florida, Oklahoma and Texas and may reach 8 feet in height. Its stems are smooth and unmarked and its leaves, from 4 to 5 inches or more in length, are arranged in whorls of five. The dark rose or purple-pink flowers are in a loose, rounded cluster, and appear in late summer.

FAIRY BELLS

TYPE Herbaceous perennial
FAMILY *Liliaceae* (Lil-ee-*ay*-see-ee)
GENUS *Disporum* (Dis-*poh*-rum)
SPECIES *lanuginosum* (lan-yew-jin-*noh*-sum)
ORIGIN Ontario to Georgia and Tennessee

There are several other species of Fairy Bells, mostly native to Western North America and all rather similar to the one described here. Fairy Bells is a close relative to the true Solomon's-seal and False Solomon's-seal, Twisted-stalk and Bead-lily, and like these other modest members of the Lily family may suffer from comparison with more flamboyant natives that flower in spring. Yet Fairy Bells and its relations have charm and perform well as background and filler plants.

USES For woodland and shady borders, in company with the above-mentioned plants as well as others like Lily-of-the-valley, Hosta, Ferns and naturalized spring-blooming bulbs. Fairy Bells is attractive among Rhododendrons and Mountain-laurel. It shows up well against a stone wall or can be spotted among rocks or against boulders in shade. In common with Solomon's-seal and the like, it is effective on shaded slopes and banks so that one can look up at its drooping flower bells. Several plants are more effective than one.

HABIT OF GROWTH The angled and branched stem rises from a slender horizontal rhizome; height, 1 to 3 feet.

FLOWERS Creamy or green-tinged bells, ¼ to ¾ inch long, hang from the leaf axils. They may be in pairs or solitary; blooming time is late spring. Each flower is followed by a red berry.

FOLIAGE Narrow, ribbed leaves, slightly hairy and arranged alternately up the stem.

ASSETS Easy to establish; pleasant subjects for shade.

FAULTS Not showy enough for those who seek obvious beauty.

CULTURE Fairy Bells may be collected from the wild in either spring or fall or ordered from wild-flower specialists, preferably for fall delivery. Plant the rhizomes horizontally an inch or two deep in acid soil that has been enriched with leafmold and/or peat moss. The situation should be shaded and the soil should not become bone-dry during drought. Division of old clumps is readily accomplished by separating the slender rhizomes in early spring or fall. Seeds can be planted in the fall; remove the surrounding pulp before sowing.

FAIRYWAND

TYPE Tuberous-rooted herbaceous perennial
FAMILY *Liliaceae* (Lil-ee-*ay*-see-ee)
GENUS *Chamaelirium* (Kam-ee-*lihr*-ee-um)
SPECIES *luteum* (*lew*-tee-um)
ORIGIN Southern Ontario to Florida, west to Arkansas and Michigan

Fairywand, as a common name, seems a happy choice for this graceful member of the Lily family. Other names are Blazing Star, Unicorn-root and Devil's-bit.

USES Suitable for colonizing in open sections of a woodland garden, in the foreground of shrub groupings and in large naturalistic rock gardens.

HABIT OF GROWTH Fairywand's root is a tuber or rhizome from which grow a low rosette of foliage and a slender, flowering stem to a height of 1 to 3 feet.

FLOWERS This plant is of special interest because male and female flowers appear on separate plants, as in the case of the Holly. The male flowers—small, white and fluffy because of their numerous yellowish stamens—are crowded together on spikes, which droop at the top. The female flowers are greenish white and packed equally close together, but their raceme does not droop. Flowering time is late spring and early summer.

FOLIAGE Lilylike leaves in a low rosette, almost ground-hugging. The largest leaves, long and flat, may be 8 inches in length.

ASSETS Fairywand should be better known. It is easy to establish in moderately rich soil and provides interest after the major spring natives have flowered.

FAULTS Not for poor, dry soil.

CULTURE In the wild, Fairywand is found in both sunny and lightly shaded situations but almost always in rich, moist soil which is acid. Wild plants can be readily lifted in early spring or in the fall; established clumps can be increased by separating the rhizomes in the fall.

FALSE LILY-OF-THE-VALLEY

TYPE Herbaceous perennial
FAMILY *Liliaceae* (Lil-ee-*ay*-see-ee)
GENUS *Maianthemum* (May-*anth*-em-um)
SPECIES *canadense* (kan-ad-*den*-see)
ORIGIN Newfoundland to North Carolina, west to South Dakota

Colonies of the False or Wild Lily-of-the-valley do have a resemblance to true Lily-of-the-valley (*Convallaria majalis*) and flower at about the same time. Also both plants are in the same family. Other common names for False Lily-of-the-valley are Canada Mayflower and Heartleaf-lily.

USES An ideal ground cover where a low, carpeting plant is needed in shade and woodlands or underneath Azaleas, Rhododendrons and other shrubs and trees. In fact, plants grow right up to tree trunks, their leaves nestling against the bark.

HABIT OF GROWTH Neat and symmetrical, each plant about 4 inches high, spreading by rhizomes.

FLOWERS The resemblance of this plant to Lily-of-the-valley ends here, for instead of bell-shaped flowers, False Lily-of-the-valley carries a short 2-inch spike of fuzzy, creamy white flowers, each tiny flower having four petals. The fuzzy effect is due to protruding stamens. Blooming time is from mid-May to early June. The flowers are followed by small berries which eventually turn red.

FOLIAGE Handsome, heart-shaped leaves about 1 to 1½ inches long, light green in color and alternately clasping the stem. Each plant has from one to three leaves.

ASSETS Easy to grow and just as easy to transplant from the wild to a garden setting. Fortunately it is in plentiful supply, but if a nearby source can't be found, plants can be readily purchased from most wild-flower nurseries.

CULTURE Give False Lily-of-the-valley a woodsy soil which, ideally, should be acid and moist. Add peat moss to the soil if it lacks humus. This plant naturally grows in woods rich in leafmold and is often found on the edges of brooks and swamps. Under cultivation it will survive in rather dry areas, provided humus has been incorporated with the soil and plants receive an occasional soaking during dry periods. When moving plants from the wild to the home garden, individual plants can be dug; or, for quicker effect, lift sods containing several plants.

FALSE SOLOMON'S-SEAL

TYPE Herbaceous perennial
FAMILY *Liliaceae* (Lil-ee-*ay*-see-ee)
GENUS *Smilacina* (Smye-las-*sye*-nuh)
SPECIES *racemosa* (ras-em-*moh*-suh), *stellata* (stel-*lay*-tah)
ORIGIN See Species, below

Other names for this genus are Solomon's-plume and False Spikenard, but False Solomon's-seal seems to persist as a common name. The main difference between this genus and "true" Solomon's-seal (*Polygonatum*) is that False Solomon's-seal carries its flowers—feathery rather than bell-shaped as are Polygonatum's—in a panicle at the end of the stem rather than from leaf axils along the stem. (Other differences found in False Solomon's-seal include the lack of pronounced scars on the rhizome and a flowering stem that goes in a slightly zigzag direction rather than being straight.) The two genera, which of course are in the same family, do resemble each other and are often found growing side by side.

USES Excellent subjects for colonizing in the woodland garden, especially along pathways or in the foreground of trees and shrubs. False Solomon's-seal makes an effective accent among Ferns and in nature often grows with the Christmas Fern (*Polystichum acrostichoides*).

HABIT OF GROWTH Graceful stems, 1 to 3 feet high, rise from creeping rhizomes.

FLOWERS They are feathery and white, arranged close together on short branches at the ends of the stems, in late spring. The showy berries, which follow, are usually speckled with dark red in the fall.

FOLIAGE Lance-shaped leaves, about 3 inches long, light green, with prominent parallel veins, fewer in number than on the leaves of Polygonatum. They are alternate.

ASSETS Easy-to-grow plants which are fairly fast-spreading. The spring flowers are pretty and the fall fruit clusters are decidedly handsome. Plants are easily lifted from the wild in spring or fall, and established clumps are readily divided. Wild-flower nurseries always list the False Solomon's-seal.

FAULTS The fruit display in fall may be of short duration. Apparently birds are attracted to the berries.

CULTURE False Solomon's-seal prefers rich, acid, woodsy soil which retains some moisture. Under such conditions—and in shade—plants spread and appear to perfection. However, they will grow quite well in dry woods. The roots should be planted horizontally a few inches below the ground level. Division is simple by cutting up the rhizome, allowing a bud or buds to be retained with each section. Seeds can also be sown in the fall to germinate in the spring.

SPECIES *Smilacina racemosa.* False Solomon's-seal. Solomon's-plume. False Spikenard. (Illustrated) Native over a large part of North America from Quebec to British Columbia, south to Georgia and Arizona. Height is about 3 feet. Fall berries are white, speckled with red.

S. stellata. Starry False Solomon's-seal. Native over much of North America, including Arizona and California. (This species is also found in Europe.) Its white flowers are slightly larger than those of *S. racemosa* but are produced more sparingly. The berries are striped black. This species prefers more moist soil conditions and is tolerant of acid or nearly neutral soil.

TWO NATIVE PLANTS FOR FULL SUN AND ACID, SANDY SOIL

Bearberry, with glossy evergreen leaves and red berries, is a trailing plant, superb on dunes, among rocks and on dry, steep slopes. These plants were grown from cuttings and make a pleasant background for Crocuses.

Hudsonia is another evergreen shrubby native which often grows in proximity to Bearberry. It has golden flowers in late spring.

MAY-APPLE NEEDS SHADE AND AMPLE SPACE

May-apple spreads in rich woodland soil to form a low canopy, its leaves hiding both the ground and its own white flowers in spring. It is best for a large wild-flower garden (left) or along roads and fences (below).

TWO WILD FLOWERS FOR MASSING

Bluets or Quaker Lady (above) produces a multitude of pretty blue and white flowers on thread-like stems. Although most at home in moist meadows, it can be moved to the home grounds to form a flowering carpet. The False Lily-of-the-valley (right) is most effective when it can spread in shade.

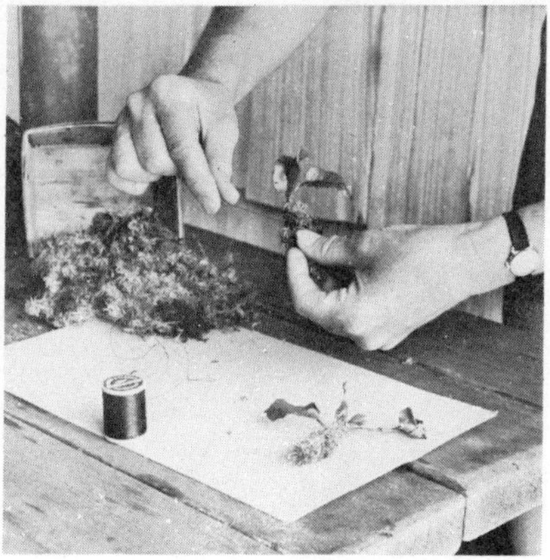

Trailing-arbutus, an evergreen sub-shrub, is difficult to re-establish when dug from the wild. A solution is to take stem cuttings in early summer, 2 to 3 inches long. Wrap lower stem with wet unshredded sphagnum moss, securing it to stem with thread.

A dozen or so cuttings are shown, all with their lower stems wrapped with sphagnum moss. When collecting the cuttings from wild plants, keep them in a plastic bag so that they don't dry out.

Fill a plant flat or any convenient wooden box with an equal mixture of sharp sand and peat moss. Firm mixture, then use blunt stick to make holes in which to insert cuttings so that sphagnum-wrapped stems are buried. Water. Cover flat with a plastic tent as shown and keep in shade.

Cuttings should be rooted in 3 or 4 weeks, the fine roots having permeated the sphagnum moss. Rooted plants, sphagnum moss and all, can be planted in an acid soil of mostly leafmold and sandy loam. Water newly set plants well, especially during drought, and mulch with Pine needles or Oak leaves over winter. Rooted plants in flower shown below.

FIREWEED

TYPE Herbaceous perennial
FAMILY *Onagraceae* (On-ag-*gray*-see-ee)
GENUS *Epilobium* (Ep-il-*loh*-bee-um)
SPECIES *angustifolium* (an-gus-tif-*foh*-lee-um)
ORIGIN North America, Europe and Asia

Most summer vacationers who drive through the Adirondacks, or through our Western national parks, or the more northerly areas of the Midwest, have been startled by the unexpected beauty of Fireweed in full flower in a roadside marsh or a burned-over, scarred area that was once a forest. The magenta flowers on tall spikes can hardly be missed when most of the surrounding vegetation is dead. This plant is called Fireweed because it so often is the most prominent plant to appear after a fire. Its fluffy seeds have the ability to fly far afield, often falling and germinating in devastated earth. Other names for Fireweed: Blooming Sally, Great Willow Herb.

USES Fireweed is suitable for accent planting by large ponds or in large swamp or bog areas or along roads or drives where a naturalized setting is being established. It makes a good companion for other giants among our natives—Ironweed, Joe-Pye Weed, the tall Asters, Cimicifuga, etc.

HABIT OF GROWTH Fireweed forms large clumps, each producing long, red-colored, flowering stems from 1 foot to as high as 7 feet.

FLOWERS They are magenta-pink or rosy-purple, 1 to 2 inches wide, arranged in long racemes or spikes like the flowers of the Delphinium. There is a rather rare white-flowered form which is very lovely. Blooming time is from early summer to early fall.

FOLIAGE Alternate, lance-shaped, olive-green leaves.

ASSETS A pretty plant in the right setting, with a long flowering period. It will grow in damp soil but is tolerant of average soil.

FAULTS Fireweed gets too tall for restricted wild-flower gardens and may become a weed.

CULTURE Grow Fireweed in full sun or light shade, in rich soil which retains moisture. It is usually found in moderately acid soils. Old clumps are easily divided in early spring. It can be propagated from seed collected in late summer from plants whose top parts are still in flower. Such seed can be planted in fall or spring.

FOAM-FLOWER

TYPE Herbaceous perennial
FAMILY *Saxifragaceae* (Sax-if-rag-*gay*-see-ee)
GENUS *Tiarella* (Tye-uh-*rell*-uh)
SPECIES *cordifolia* (kor-dif-*foh*-lee-uh)
ORIGIN Nova Scotia to Georgia and Alabama

One of our best natives for naturalizing in woodlands, under trees and around shrubs is the Foam-flower or False-miterwort. It is a close relative of both Bishop's-cap and Alum-root, resembling both of these in foliage and flower formation, although its flowers are considerably showier.

USES Allow Foam-flower to form colonies in the woodland under trees and along paths. Set plants against old logs and boulders, in rich soil pockets in a shady rock garden or at the base of a stone wall. It is especially attractive with Ferns, Primroses, and a few clumps of naturalized, late-flowering Daffodils.

HABIT OF GROWTH Several flowering stems, 6 to 12 inches high, rise from high rosettes or clumps of foliage.

FLOWERS Tiny, fuzzy flowers, white and about ¼ inch wide, are carried in a spike or raceme, from 1 to 4 inches long and resembling a cone-shaped bottle brush. Blooming time is spring, but some flowers appear in early summer.

FOLIAGE Rough-textured, heart-shaped leaves, lobed and toothed, about 2 to 4 inches long.

ASSETS Its tolerance of shade, its dainty flower spikes and attractive foliage. Foam-flower is easy to grow and propagate.

CULTURE Rich woodland soil that is neutral or moderately acid suits Foam-flower, and it needs a shady exposure. Before planting, incorporate quantities of peat moss and/or leafmold into the soil. It will do fairly well in dry woodlands if the light, sandy soil is reinforced with a mixture of clayey soil as well as the humusy materials just mentioned. In damp, rich situations, Foam-flower sends out many runners, which, when rooted, can be easily cut from the parent and replanted.

FRINGED ORCHID

TYPE Herbaceous perennial
FAMILY *Orchidaceae* (Or-kid-*day*-see-ee)
GENUS *Habenaria* (Hab-en-*nay*-ree-uh)
SPECIES *ciliaris* (sil-ee-*ay*-riss), *fimbriata* (fim-bree-*ay*-tuh), *psycodes* (sye-*koh*-deez)
ORIGIN See Species, below

The Fringed Orchis or Orchid is a true Orchid and somewhat resembles some of the exotic, small-flowered Orchids of the tropics. Another name is Rein Orchid, the "rein" referring to the long spur found in the flowers of some species. There are many species of Habenaria—perhaps 500 species throughout the world—most of which are considered too chancy to cultivate. The ones listed here are recommended for the home wild garden and are sold by some wild-flower nurseries. Other native species described in wild-flower manuals are worth transplanting if found in quantity and if their wild growing conditions can be duplicated.

USES For either natural or man-made bog areas or moist, open woodlands, grown in groups with Ferns and against weathered logs. Also in rich soil pockets in a shaded rock garden.

HABIT OF GROWTH Upright, tuberous-rooted clumps, some low, others fairly tall and vigorous.

FLOWERS The individual flowers, though small, are exquisite, and are in racemes or spikes from barely a foot to 2 to 3 feet, depending on the species and its growing conditions. The flowers of the species listed here have a prominent lip, heavily fringed, with a nectar-producing spur 1 inch or more in length. The color range of this genus is wide, ranging from white to greenish white, pink and purple and bright gold or orange. In some species the flowers are scented.

FOLIAGE Usually the oval or lance-shaped leaves, few in number, are ground-hugging, with much smaller, even scale-like, leaves on the flowering stems. Size varies according to the species.

ASSETS The beauty and variety of the flowers, which alone is reason enough to grow them. A second attribute is that they bloom in summer and late fall when few other native plants of great quality are in flower. Thirdly, the flowers are long-lasting; even the ball-like bud stage of *Habenaria ciliaris* is effective for several days before the golden, fringed flowers reach full development.

FAULTS Not all Fringed Orchids are easy to grow. As is the problem with all fleshy-rooted plants whose roots are near the surface, they may be destroyed by rodents.

CULTURE The species listed here generally like moist, acid soil, heavy in humus, although *H. ciliaris* is often found in quite dry situations. Mix leafmold or peat moss and sand in their planting area, and set the tuberous roots horizontally, just below the surface. Plants are available from native plant nurseries in spring and fall. Collecting plants from the wild is recommended only when they are plentiful and in late fall or early spring, and, as with any demanding native, the habitat should be noted so that the uprooted plants can be replanted in conditions as similar as possible to those in the original site. Additional growing suggestions are given below.

SPECIES *Habenaria ciliaris.* Yellow Fringed Orchid. Orange Fringed Orchid. This Orchid almost belongs to the "easy" class because it is so widely distributed in such a variety of situations. It is found in the East from New England down to southern Florida; along the Gulf Coast to Texas; in Missouri and Arkansas and northward to Michigan, Wisconsin and Illinois. The degree of acidity in the soil varies from very acid to near neutral; the soil may be rich in humus, as it is in bogs and swamps and wet meadows, or rather dry and sandy as in some Pine barrens. Plants are also found in both shady and sunny situations. Although this gives the home gardener wide leeway, good sense dictates that this Fringed Orchid be planted in acid, rather sandy soil full of leafmold and/or peat moss, probably in light shade. The strongly fringed, bright flowers appear in midsummer and are one of the highlights of the summer season among native plants. They are especially long-lasting. The flower stem, from 12 to 30 inches tall, rises above a rosette of lance-shaped leaves. *H. fimbriata.* Purple Fringed Orchid. (Illustrated) Native to northeastern North America, as far south as the mountains of North Carolina and Tennessee and west to Michigan, Illinois and Wisconsin. This very beautiful native is considered by some to be a larger-flowered variety of *H. psycodes.* Its fragrant, mauve-pink, heavily fringed flowers are crowded on stems up to 24 inches or more in height and appear in early summer to midsummer. Each clump has several narrow, pointed leaves, about 9 inches in length, with many shorter, narrower leaves on the flowering stem. Grow in acid, rich moist soil enriched with leafmold or peat moss in light shade or sun. This species needs cool conditions in summer. If grown in full sun, be sure some leafy plant is nearby to shade the roots.

H. psycodes. Small Purple Fringed Orchids. Native from Newfoundland south to Georgia and Tennessee and west to Minnesota. A gem, one of the most exciting of all natives, and fortunately, considered fairly easy to cultivate. It has fragrant, purple flowers in midsummer, which when compared with *H. fimbriata* form a narrower raceme. Its flowers are also heavily fringed, the lip being divided into three parts. Grow in moist, humus-rich soil in light shade.

GALAX

TYPE Evergreen perennial
FAMILY *Diapensiaceae* (Dye-ap-pen-see-*ay*-see-ee)
GENUS *Galax* (*Gay*-lax)
SPECIES *aphylla* (af-*fill*-uh)
ORIGIN Virginia and West Virginia to Georgia and Alabama

Galax and its near relative, Oconee-bells, are two of the most distinctive ground covers available for Northern woodland gardens, even though they originate in Southern mountains. Galax, sometimes called Galaxy or Beetleweed, has handsome evergreen leaves which are collected in great quantity and then shipped to florists all over the country, mostly for use in funeral sprays. When bouquets of Violets were in style, Galax leaves were used as an outer decoration. Galax deserves a better fate, and fortunately the plants are not hard to establish in shaded, acid-soil gardens.

USES As ground cover and accent in woodland gardens, shaded rockeries, along paths or as underplantings to naturalized groupings of Rhododendrons, Azaleas, Dogwoods and other trees and shrubs.

HABIT OF GROWTH Low-growing, stemless plants forming leafy clumps up to 6 inches or so high from underground runners.

FLOWERS Small white flowers arranged in a spike 10 to 20 inches above the leafy base in late spring.

FOLIAGE Beautiful, heart-shaped leaves from 2 to 5 inches across, of a leathery texture, the veins prominent on the waxy surface. The edges are attractively scalloped. In fall the leaf margins turn red, and leaves in exposed situations in winter become bronze.

ASSETS The evergreen leaves on their stiff stems are always attractive, and plant colonies have added charm in spring when accented by the spikes of white flowers. Plants are tolerant of deep shade.

FAULTS Galax is another native which must have acid soil.

CULTURE Most wild-flower nurseries offer plants of Galax, which are shipped in early spring or fall. A few plants set out about 6 inches or so apart will spread by their underground runners to form extensive colonies. Plants can be divided in early fall, each potential clump possessing a single white bud at the soil surface or just beneath it. Select a location in full or partial shade and mix quantities of leafmold or peat moss in the soil, which should remain uniformly moist but not waterlogged.

GAYFEATHER

TYPE Herbaceous perennial
FAMILY *Compositae* (Kom-*poz*-it-ee)
GENUS *Liatris* (Lye-*ay*-triss)
SPECIES *ligulistylis* (lig-yew-liss-*tye*-liss), *pycnostachya* (pik-noh-*stak*-ee-ah), *scariosa* (skay-ree-*oh*-suh), *spicata* (spye-*kay*-tuh)
ORIGIN See Species, below

How did Liatris, with royal-purple, Thistle-like flowers, ever get the common name of Blazing Star? A more unsuitable name is hard to imagine. Far more appropriate is Gayfeather. These pretty perennials have long been prized for flower gardens and are among the North American natives which were introduced early to Europe.

USES Around pools and ponds, either man-made or natural, in perennial borders, or naturalized or grouped in meadows or at the edge of a woodland in any planting of such late summer wild flowers as Goldenrod, Golden-aster, Eupatorium, and Aster. All species of Gayfeather make long-lasting cut flowers, so it pays to grow as many as possible for this purpose.

HABIT OF GROWTH Stiffly upright in clumps from a tuberous, almost bulblike root, each root usually sending up from one to several stems, generally unbranched.

FLOWERS Rich purple, sometimes with a rosy cast, flowers in Thistle-like or button heads along the stiff stems, each stem carrying a number of flower heads (up to 40 or more on some species). The terminal flowers open first.

FOLIAGE Narrow, bladelike leaves, alternate, and resembling those of some Lilies in their arrangement and density on the stems.

ASSETS The long-lasting, bright purple flowers are especially appropriate to late summer and early fall, and are attractive in groupings by themselves or used as contrast and accent among other perennials. Plants of several species are available from most wild-flower specialists and improved forms have been introduced. They are all easy to grow; some are especially tolerant of dry soil.

FAULTS Species of Gayfeather have hybridized freely and even the species themselves do not differ very much, so it is trying to all except a botanist to sort them out.

CULTURE The Gayfeathers like full sun but will tolerate some shade during the day. Plant in spring or fall, making a grouping of three or more tubers to achieve a more colorful display. Species such as *L. ligulistylis, L. pycnostachya,* and *L. spicata* need a moderately rich, damp soil, as they naturally inhabit moist meadowland. *Liatris scariosa* is usually abundant in dry, sandy soil, so the same conditions can be duplicated in the home garden. Generally all the Gayfeathers seem to be tolerant of good average garden soil with average conditions. Propagation is by seed, best sown in the fall, or by dividing the clumps in early spring. Plants can be collected from the wild in early spring (mark their location when they flower in summer) or fall.

SPECIES *Liatris ligulistylis.* Native from Manitoba to Colorado and adjacent areas. Purple flowers, 1 inch across on spikes about 18 inches high, in July. Grow in rich soil and light shade.

L. pycnostachya. Prairie Gayfeather. Cat-tail Gayfeather. Native from Minnesota to Louisiana and Texas. Purple flowers on 3- to 5-foot stalks in late summer with alternate, grassy leaves.

L. scariosa. Large Button Snakeroot (Illustrated) Maine to northern Florida and Texas. Rosy-purple flowers on 3- to 5-foot stalks. Grow in dry soil, full sun.

L. spicata. Spike Gayfeather. Native from Massachusetts to Florida and Louisiana. Purple flower heads on 3- to 5-foot stalks in August. Grow in rich, moist soil in full sun.

GENTIAN

TYPE Herbaceous biennial and perennial
FAMILY *Gentianaceae* (Jen-shee-uh-*nay*-see-ee),
GENUS *Gentiana* (Jen-shee-*ay*-nuh)
SPECIES *andrewsi* (an-*drew*-sye), *crinita* (krih-*nye*-tuh), *decora* (*dek*-or-uh), *saponaria* (sap-on-*nay*-ree-uh)
ORIGIN See Species, below

Our native Gentians, especially the electrifying Fringed Gentian (*Gentiana crinita*), are to autumn what Trailing-arbutus (*Epigaea repens*) is to spring. Both the Fringed Gentian and Trailing-arbutus can be tamed, but such a task is not always easy, and for many, enjoyment of these plants is only possible by observing them in their native habitat. Fortunately among the Gentians there are species which are nearly as lovely as the Fringed Gentian and which can be grown quite readily in the wild-flower garden. There are many other Gentian species native to North America, many of which are quite similar to the species considered here. These four are sold by specialists, either as seeds or plants.

USES The native Gentians like damp soil. Grow them in bogs, along brooks or around informal pools or ponds or in rich soil that never becomes bone-dry.

HABIT OF GROWTH Upright or sprawling clumps, about 2 feet tall.

FLOWERS Gentians have bright lilac-blue flowers, the intensity of the blue coming through more brightly than the degree of lilac. While the color of a Gentian is hard to describe, its effect upon the viewer is startling, especially when a whole meadow of the plants is encountered. The species described here have fringed petals, carried in a graceful, elongated bud which unfurls for only a few hours before closing again, or oval, bottle-shaped buds which never open. Flowering time is fall.

FOLIAGE Opposite, pointed leaves, their size and color varying with the species.

ASSETS The brilliance and beautiful form of the flowers place the Gentians high in the esteem of plant lovers. For the Fringed Gentian, especially, almost any means is worth while to bring the plants into flower in the home garden. The other species are quite easy to establish either as purchased or collected plants.

All bloom over a long period in the fall, providing interest in the wild-flower garden long after many other natives have finished flowering.

FAULTS A biennial is grown from seed one year, blooms the second year and then usually dies. The Fringed Gentian is a biennial. It must be transplanted with great care, only in its early seedling stages.

CULTURE All these native Gentians like rich, moist soil and are often found growing in boggy meadows or wet ditches difficult to duplicate in the home wild-flower garden. Fortunately a rich, humusy soil (with some clay plus a great deal of leafmold and/or peat moss) which remains moist will suit them. Give them full sun or at least part sun. The perennial Gentians listed here can be planted in spring, and established clumps can be readily divided at the same season. The Fringed Gentian, a biennial, can be grown from seed sown in late fall (fresh seed gives the best germination) in a pot or flat of peat moss and sand which should be left outdoors all winter. The seeds will germinate in the spring and form tiny rosettes, at which time they can be gingerly transplanted into their permanent positions in well-prepared, humus-rich soil that will stay moist. The seedlings should be shaded until they become established. For those who inherit a moist meadow, or have some other site that seems suitable, chances are good that Fringed Gentian seed sown directly in the area in the fall will take hold. In fact, once it is established in the right environs, the Fringed Gentian may self-sow. The "may" is tentative, as the Fringed Gentian is temperamental. When nature is the gardener, it may grow with such abandon as to fill meadows and roadsides; when man tries, it may refuse to cooperate!

SPECIES *Gentiana andrewsi.* Closed or Bottle Gentian. (Illustrated, left) Native over Eastern North America. The oval, pointed flowers of bright blue are produced in clusters at the leaf axils and are nearly as showy as the flowers of the Fringed Gentian. The leaves, dark green and lance-shaped, are about 3 inches long. This species, perennial and easy to grow in moist, humusy soil, is offered by all wild-flower specialists.

G. crinita. Fringed Gentian. (Illustrated, right) Eastern North America. Patches of blue made by this aristocrat were once an expected glory of late autumn, but spreading suburbs and vast roadway development as well as "clean" cultivation along even country roads mean that one must search deeper and farther for Gentian fields and colonies. Thankfully, such plantations still exist. It is perfectly correct to collect seed late in the fall from such plantations—to be planted in protected meadowlands or introduced into the suitable home landscape, even if only on a small scale.

G. decora. Native to the Alleghenies. Similar to the Bottle Gentian, although the leaves are narrower and the buds show a slight fringing of their petal tips.

G. saponaria. Soapwort Gentian. Also similar to the Bottle Gentian.

GINSENG

TYPE Herbaceous perennial
FAMILY *Araliaceae* (Ar-ray-lee-*ay*-see-ee)
GENUS *Panax* (*Pay*-nax)
SPECIES *quinquefolium* (kwin-kwef-*foh*-lee-um), *trifolium* (trye-*foh*-lee-um)
ORIGIN See Species, below

Not too many years ago many American seed catalogs featured special offers of Ginseng seed, with instructions on how to grow it as a commercial crop for its tuberous roots which were to be shipped to China. The Chinese believed they possessed aphrodisiac as well as other medicinal values—apparently a doubtful assumption—but nevertheless Ginseng was grown and collected from the wild for this purpose and now has been wiped out of many of the areas where it once flourished in great numbers. Actually it is not a very attractive plant, although Dwarf Ginseng (*P. trifolium*), which is illustrated, is quite pretty and is still fairly common in rich woods.

USES Both species are shade lovers and should be grown in a woodland garden. Dwarf Ginseng can be planted in rich soil pockets in a naturalistic and shaded rock garden or on shaded banks or slopes.

HABIT OF GROWTH Both Ginseng (*P. quinquefolium*) and Dwarf Ginseng have tuberous roots which go deep into the ground. From the root rises a single stem, which in the case of Ginseng is from 1 to 2 feet high and forms three short branches at its top for the foliage. Dwarf Ginseng grows from 3 to 8 inches high.

FLOWERS Ginseng produces an umbel of greenish flowers in late spring and early summer which is followed by a cluster of red berries in early fall. Dwarf Ginseng has much prettier flowers, sparkling white or pinkish, in a lacy cluster usually produced in mid-spring. Its berries are yellow.

FOLIAGE Deep green, compound leaves on a long petiole divided palmately into five segments (in the case of Ginseng), each leaflet from 2 to 5 inches long. The compound leaves of Dwarf Ginseng are divided into threes usually, sometimes fives, and are carried on much shorter petioles than are those of Ginseng. Each leaflet is about 2 inches long and has toothed margins.

ASSETS The rather stiff, symmetrical form of Ginseng can make an interesting accent in the woodland garden. Very desirable is Dwarf Ginseng for its perky charm when flowering in early to mid-spring. A colony can be most effective.

FAULTS Ginseng's flowers possess no special beauty. Both species are difficult to dig successfully.

CULTURE Grow the Ginsengs in shade, in humus-rich soil which will retain moisture in periods of drought. Most wild-flower nurseries offer plants of both species which are usually shipped in fall, the preferred planting time. The plants can be propagated from seed, to be sown in the fall for spring germination.

SPECIES *Panax quinquefolium.* Ginseng. True Ginseng. Native from Quebec to Minnesota and Nebraska, south to Georgia, Ohio and Iowa. This rather tall, coarse native is worth growing in a large woodland for accent purposes and possibly as a conservation measure, since it has disappeared from so many areas. *P. trifolium.* Dwarf Ginseng. Groundnut. The common name Groundnut refers to its dark, rounded, tuberous root. By far the more useful and desirable species.

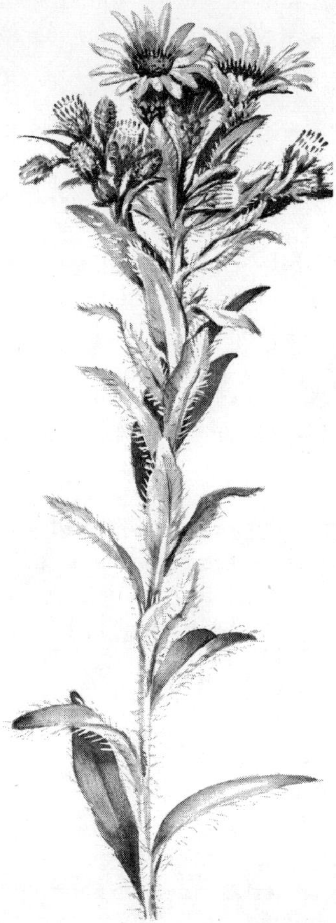

GOLDEN-ASTER

TYPE Herbaceous perennial
FAMILY *Compositae* (Kom-*poz*-it-ee)
GENUS *Chrysopsis* (Kriss-*op*-siss)
SPECIES *falcata* (fal-*kay*-tuh), *mariana* (mar-ee-*ay*-nuh), *villosa* (vil-*loh*-suh)
ORIGIN See Species, below

This golden Daisy of late summer is found in dry, open woodland and meadows and is well worth transferring into the wild-flower garden.

USES It should be grown in colonies in woodlands, naturalistic rock gardens, along paths or on dry slopes or banks. It makes a good companion for Goldenrod, Aster and Gayfeather.

HABIT OF GROWTH Upright clumps, sometimes matlike at the base, with one or several flowering stalks, covered with white hairs, up to about 18 inches.

FLOWERS Daisies, from ¾ to 1 inch across, bright yellow, and carried in loose, irregular heads at the ends of the stalks, in late summer.

FOLIAGE Alternate, hairy, lance-shaped leaves about 2 inches long, larger at the base, clasping the stems at close intervals.

ASSETS The bright yellow flowers can add great cheer to an otherwise drab open spot of woodland in late summer. Clumps add perky interest when established in rock outcroppings or nestled at the foot of boulders or in soil pockets of sunny rock gardens. They are drought-resistant, and along the seashore they will tolerate sand, sun and wind.

CULTURE A dry, lean acid soil seems to suit them. Add leaf-mold or peat moss to very sandy soil before planting. Established clumps will soon self-sow freely. Old clumps can be divided in the spring. Clumps can be dug in full flower, but should be watered for the first few days after transplanting. Plants can be cut back in June for a more compact habit.

SPECIES *Chrysopsis falcata.* Golden-aster. Found in dry, sandy soil from Massachusetts to New Jersey, especially along the coast. The flowers, bright yellow Daisies, are ¾ inch across, and appear from midsummer into fall. The narrow, light green leaves may be twisted or sickle-shaped. Height is about 1 foot. An ideal species for dune cover.

C. mariana. Maryland Golden-aster. Broad-leaf Golden-aster. (Illustrated) Native from southern New York to Pennsylvania, Florida and Louisiana. Bright yellow Daisies, 1 to 1½ inches across, from midsummer to early fall. Height is about 2 feet.

C. villosa. Midland Golden-aster. Native from Minnesota to British Columbia and New Mexico. This more widely distributed species has slightly larger flowers and is more compact in its growth habit.

GOLDEN RAGWORT

TYPE Herbaceous perennial
FAMILY *Compositae* (Kom-*poz*-it-ee)
GENUS *Senecio* (Sen-*nee*-see-oh)
SPECIES *aureus* (*aw*-ree-us)
ORIGIN Labrador to North Dakota, south to Florida, Alabama, and Arkansas

This hardy, Daisy-flowered perennial which flowers in spring is in the same genus as such tender, exotic plants as German-ivy, Cineraria, and the beautiful gray-leaved Dusty Miller. The Golden Ragwort or Golden Groundsel is one of several other species native to North America and has been selected for its early blooming habit and general availability from wild-flower specialists.

USES For naturalizing in rich woodlands or in swamps or around good-sized pools or ponds.

HABIT OF GROWTH Upright, branching plants from 1 to 3 feet tall, with a leafy base.

FLOWERS Golden Daisy flowers, ½ to ¾ inch across, in loose, flat-topped clusters, from mid- to late spring, some flowers usually appearing well into summer.

FOLIAGE The Golden Ragwort has two types of foliage; its basal leaves are decidedly heart-shaped, about 1 to 6 inches long, and have long petioles. The flowering stem leaves are clasping, elongated and finely cut.

ASSETS Attractive, bright flowers with a long blooming period, and interesting foliage. The Golden Ragwort is a fine choice for damp, poorly drained soil.

FAULTS Not for dry soil.

CULTURE Plant clumps of the Golden Ragwort in spring or fall in neutral or moderately acid soil in part shade or sun. The soil should be rich in humus and if not naturally wet, must retain moisture so that the plants never become bone-dry. Division is easy in early spring. Established plants readily self-sow.

GOLDENROD

TYPE Herbaceous perennial
FAMILY *Compositae* (Kom-*poz*-it-ee)
GENUS *Solidago* (Sol-id-*day*-goh)
SPECIES *canadensis* (kan-ad-*den*-siss), *cutleri* (*cut*-ler-eye), *mollis* (*moll*-iss), *nemoralis* (nem-or-*ray*-liss), *odora* (oh-*doh*-ruh)
ORIGIN See Species, below

The Goldenrods do not hoard their gold but spill it bountifully over fertile meadows and waste areas alike. Goldenrod is a major contributor to the rich color effects of late summer and autumn landscapes. This genus has a horde of species, and sorting them out is a project for botanists. Gardeners can look around their regions and then decide which Goldenrods are worthy of special notice and which ones should be left to nature. (The Goldenrod was once despised for being the cause of hay fever, but the villain is Ragweed.)

USES Most Goldenrods are sun lovers although some, like the Anise-scented Goldenrod (*S. odora*), are pretty in patches in open woodland. A few, mostly those native to mountains, are compact enough for rockeries and retaining-wall use. Most are best in informal sunny borders in meadows, at the edge of woodlands, along drives and, in short, where nature put them in the first place.

HABIT OF GROWTH Clumpy, upright plants, the flowering stems varying from a few inches in height to several feet.

FLOWERS The small ray-and-disk flowers of the species mentioned here are yellow or gold (a few other species have nearly white or pale yellow flowers) and are packed close together in panicles or racemes. Blooming time is from summer to fall.

FOLIAGE There is great variation but most Goldenrods have many leaves, sometimes handsome, sometimes coarse. Those with gray leaves can be very attractive.

ASSETS The Goldenrods are easy to grow and tolerate poor soil. The flowers are especially long-lasting and some species are in flower from midsummer until late fall.

FAULTS Most Goldenrods perfectly fit the description of a weed: a plant out of place. They self-sow, but seedlings and mature plants alike are easy to pull up by hand in small gardens.

CULTURE The Goldenrods listed here—and most other species, for that matter—grow in average soil, even in soil that is dry and poor, and need full sun or some sun for part of the day. Plants can be transplanted, the best time being in spring. Old clumps can be divided.

SPECIES *Solidago canadensis.* Canada Goldenrod. Native from Newfoundland to Florida, west to Arizona, South Dakota and British Columbia. This widely distributed Goldenrod with yellow flowers in one-sided racemes on tall stems, 5 to 7 feet high, begins to bloom in late summer and reaches its peak in the fall. It needs full sun and is too tall for all but special situations such as along fences or at the edge of woods.

S. cutleri. Alpine Goldenrod. Native to mountains of New England and New York. A fairly compact species, from a few inches to 12 inches in height, its flowers in short racemes in late summer. Useful in sunny rockeries and in retaining walls.

S. mollis. Hoary Goldenrod. Velvety Goldenrod. Midwest species with distribution to British Columbia and south to Texas and Mexico. Another compact species with attractive, stiff gray foliage and soft gray hairs on short stems from 4 to 12 inches high. The especially bright gold flowers appear in early fall. A showy, non-weedy Goldenrod for dry soil and sun, especially around rocks.

S. nemoralis. Gray Goldenrod. Native over wide area: Nova Scotia to Florida, west to Saskatchewan, Arizona and Texas. Golden-yellow flowers in one-sided curved panicles, the flowering stems from 12 to 24 inches high. The leaves and stems are gray-green. This Goldenrod, though common, is attractive in sunny situations with poor, dry soil. Its flowering period is long, from midsummer to late fall.

S. odora. Anise-scented Goldenrod. Fragrant Goldenrod. Nova Scotia to Florida, west to Missouri and Texas. Small, yellow flowers in summer, its thin leaves giving a pleasing Licorice or Anise scent when crushed. Over-all height of the plant is 2 to 4 feet. Suitable for open, dry woodland.

GOLDENSEAL

TYPE Herbaceous perennial
FAMILY *Ranunculaceae* (Ran-nun-kew-*lay*-see-ee)
GENUS *Hydrastis* (Hye-*drass*-tiss)
SPECIES *canadensis* (kan-ad-*den*-siss)
ORIGIN Vermont to Minnesota and Nebraska, south to Georgia, Alabama and Arkansas

Conservationists are concerned about the scarcity of Goldenseal, which has been ruthlessly collected from the wild for the alkaloid hydrastine, which is derived from its roots for medicinal use. Home gardeners can help save Goldenseal by buying plants from wild-flower specialty nurseries. Other common names: Orange-root, Ground-raspberry.

USES As colonies or accents in woodlands and along shaded paths.

HABIT OF GROWTH It is intriguing in that from its orange-yellow, fleshy root comes a large leaf, then a stem from 10 to 20 inches high with two more leaves near the single flower. After fruiting, the aboveground parts generally die down.

FLOWERS Only one flower, greenish-white, about ⅓ inch across, is produced by each stem, just over the pair of leaves mentioned above. The flower is not very effective but the red Raspberry-like fruit which follows is showy.

FOLIAGE Handsome and distinctive. The basal leaf is toothed and palmate, 5 to 8 inches across, and carried on a long petiole. The stem leaves which appear below the flower are similar but smaller.

ASSETS An effective accent with its lush foliage and bright fruit.

FAULTS Foliage may disappear after the fruit forms when the season is especially dry.

CULTURE Grow in woodland in soil rich in leafmold and moderately moist. Rhizomes of established plants can be divided in spring or fall.

GOLDEN-STAR

TYPE Herbaceous perennial
FAMILY *Compositae* (Kom-*poz*-it-ee)
GENUS *Chrysogonum* (Kriss-*og*-on-um)
SPECIES *virginianum* (vir-jin-ee-*ay*-num)
ORIGIN Southern Pennsylvania to Florida and Alabama

This sunny, easygoing native deserves wider planting. Its golden flowers can be a pleasant surprise in spring woodlands, especially near naturalized Daffodils and a few clumps of Virginia-bluebells or Jacob's-ladder.

USES Most attractive in rock gardens or in soil pockets in rock ledges located in sun or partial shade, as colonies in an open woodland, on banks or slopes, or along walks and at the base of stone walls. It is effective with spring bulbs, especially Grape-hyacinths (*Muscari*).

HABIT OF GROWTH A leafy, clumpy plant which spreads by runners. Its first flowering stems are usually about 6 inches high but by summer may reach 12 inches or more.

FLOWERS Star-shaped Daisies, golden yellow and about 1 to 1½ inches wide, produced quite freely from mid-spring into summer.

FOLIAGE Clumps of hairy, toothed leaves which continue up the flowering stalks in an opposite arrangement.

ASSETS Bright flowers produced over a long period, tolerance of part shade or full sun and no special soil requirements.

CULTURE Dry woodland soil or that of the average flower garden seems to suit Golden-star equally well, and while it performs adequately in sun, it prefers light shade and puts on its best display near and around rocks and boulders. It is easily increased in early spring by lifting the clumps and pulling apart rooted sections.

GOLDTHREAD

TYPE Herbaceous or semi-evergreen perennial
FAMILY *Ranunculaceae* (Ran-nun-kew-*lay*-see-ee)
GENUS *Coptis* (*Kop*-tiss)
SPECIES *groenlandica* (gro-en-*lan*-di-cuh)
ORIGIN Greenland, to Alaska, south to North Carolina, Indiana and Iowa

An examination of the roots, before planting this little native, will show why it is called Goldthread. This species is still listed as *Coptis trifolia* by many catalogs. That species is native to Asia.

USES A delicate ground cover for damp, acid woodland; or better yet, introduce Goldthread around the fringes of a bog garden, man-made or natural.

HABIT OF GROWTH Creeping by means of its spreading, golden, threadlike roots. Its leaves barely rise above the earth; its flowering stems may reach 6 inches.

FLOWERS Buttercup-like white flowers produced singly on slender, 4- to 6-inch stems, in spring, some flowers appearing in early summer.

FOLIAGE Leathery, three-parted leaves, 1 to 2 inches across, on long petioles. Usually some leaves remain evergreen if protected over winter by leaves and debris.

ASSETS An attractive ground cover for shaded, moist, cool conditions.

FAULTS Goldthread is not suited to dry woodlands with poor soil.

CULTURE Many nurseries offer "sods" or clumps of Goldthread which should be planted in moist, acid soil full of humus. Add quantities of leafmold or peat moss to the planting holes. When planted in a bog garden, it will tolerate some sun. Plants are easily lifted from the wild, and established clumps can be divided. In nature, Goldthread is often found growing with the False Lily-of-the-valley, *Maianthemum canadense*. This combination is easily duplicated when Goldthread is introduced into the wild-flower garden.

GRASS-OF-PARNASSUS

TYPE Herbaceous perennial
FAMILY *Saxifragaceae* (Sax-if-rag-*gay*-see-ee)
GENUS *Parnassia* (Par-*nass*-ee-uh)
SPECIES *caroliniana* (ka-rol-in-ee-*ay*-nuh), *fimbriata* (fim-bree-*ay*-tuh), *palustris* (pal-*lust*-riss)
ORIGIN See Species, below

Not Grasslike at all is this late-summer charmer, although its slender flowering stems do give an illusion of Grass.

USES Strictly wet-soil plants, and in the wild-flower garden they are perfectly suited to pool and pond side. Other suitable situations include natural or man-made bogs, wet meadows or the foreground of borders containing Cardinal-flower, Gentian, Primrose and other plants requiring damp soil.

HABIT OF GROWTH From a thick root rises a low tuft of foliage, about 4 inches tall; the flowering stalk is 6 to 18 inches or more high.

FLOWERS Each flower, borne singly at the top of the straight stem, is exquisite in form, possessing five white petals delicately veined in green. An established clump may produce from twelve to fifteen flowers at a time, and as these may be up to 1½ inches across and are long-lasting, their effect is decidedly handsome. Blooming time is late summer and early fall.

FOLIAGE Heart-shaped leaves, usually semi-evergreen, 1 to 2 inches long, clasping the stem or at the base of the clump on long, 2- to 4-inch petioles. Each leaf has palmate veinings.

ASSETS A charming accent for the edges of naturalistic pools or for any moist-soil situations. Its flowers are a relief from the more flamboyant effects expected in late summer and fall.

FAULTS The Grass-of-Parnassus must have wet soil, which limits its use in the average wild-flower garden.

CULTURE Grow the Grass-of-Parnassus in sunny bogs, wet meadows or at the edge of pools where, if the soil isn't naturally wet, the plants will at least have the benefit of some overflow from the pool. The soil, in addition to being moist or even wet, must have an abundance of humus in the form of leafmold or peat moss. Experts differ on the chemical value of soil for Grass-of-Parnassus. In nature, plants are found growing in the gravelly soil along streams which may not necessarily be acid, or they may be found in limestone outcroppings which are only slightly acid or neutral. In known acid-soil areas, a sprinkling of ground limestone (not hydrated lime) might be mixed with the soil before planting Grass-of-Parnassus. It should have sun or filtered sunlight. Most nurseries offer plants for spring or fall planting. Established clumps can be carefully divided in spring.

SPECIES *Parnassia caroliniana.* Grass-of-Parnassus. (Illustrated) Native from New Brunswick to Manitoba, south to Virginia, west to Illinois and Iowa. This species is sometimes listed as *P. glauca.* The several species of Parnassus found in North America are very similar. Height, about 20 inches, flowering time late summer and fall.

P. fimbriata. Native from Colorado and California to Alaska. This species, found along mountain brooks, has fringed petals.

P. palustris. Widely distributed in North America and Europe. Very similar to *P. caroliniana.*

GRASS-PINK ORCHID

TYPE Herbaceous perennial
FAMILY *Orchidaceae* (Or-kid-*day*-see-ee)
GENUS *Calopogon* (Kal-oh-*poh*-gon)
SPECIES *pulchellus* (pull-*kell*-us)
ORIGIN Newfoundland to Ontario and Minnesota, south to Florida and Missouri

The specific name of this Orchid means "little beauty," which perfectly describes the lovely pink flower. The Grass-pink Orchid was one of the first Orchids to be sent to England; records indicate that it was being cultivated there before 1771. The enterprising Dutch were growing it before 1740.

USES For woodland gardens, rockeries and bogs and pool margins in sun or light shade.

HABIT OF GROWTH A slender flowering raceme grows from a bulblike corm with several fleshy, horizontal roots. Height of the flowering stem is from 6 inches to a couple of feet.

FLOWERS Each flower, about 1 inch across and rose-pink or magenta, has a prominent lip ½ inch long, covered with bright hairs which may be orange or purple. There may be from three to as many as fifteen flowers to a stem. The flowers, lightly scented, appear in early summer and summer. Plants with white flowers have been reported.

FOLIAGE There is a single, narrow, ribbed leaf, shorter than the flowering stem.

ASSETS A lovely Orchid, quite adaptable and not impossible to grow for those who can give it damp, acid soil.

FAULTS Obviously not for everyone. There is no sense in trying to grow this Orchid in dry, non-acid soil.

CULTURE Sun or light shade are suitable. This is a plant for very acid soil that stays damp or at least retains some moisture. Some growers recommend planting the Grass-pink Orchid in sphagnum moss, either in a bog or mixed, chopped, into the planting mixture. The soil should be quite sandy but should contain humus, so incorporate leafmold and/or peat moss when planting. The Grass-pink Orchid is often found in sphagnum bogs, Pine barrens and wet meadows, so something approximating these situations should be provided in the home garden. Unfortunately, the Grass-pink Orchid is a favorite of burrowing mice, so planting the corms in a wire basket might be worth while where practical.

HEPATICA

TYPE Evergreen perennial
FAMILY *Ranunculaceae* (Ran-nun-kew-*lay*-see-ee)
GENUS *Hepatica* (He-*pat*-i-kuh)
SPECIES *acutiloba* (ak-kew-til-*loh*-buh), *americana* (am-eh-rik-*kay*-nuh)
ORIGIN See Species, below

The beloved Hepatica or Liverleaf has always been a favorite for early spring adventurers into the woods. Some years it may flower as early as March in the vicinity of New York, and along with the bright green of Skunk-cabbage it is one of the few signs that spring is actually on the way. Hepatica is in the Buttercup family, a kinship fairly obvious when its lobed leaves and flower forms are observed.

USES Suitable for colonizing in woodland gardens or in any shaded area such as informal retaining walls or rock gardens.

HABIT OF GROWTH Each plant forms a tuft or stemless clump, both the leaves and flowers rising 6 inches from the crown. The spring growth, including new leaves, stems and flower buds, is covered with fine hairs.

FLOWERS Pretty flowers with petal-like sepals, about ½ inch across, varying in color from white to pink and shades of blue and lavender.

FOLIAGE Each leaf has three lobes and is from 1 to 2½ inches across. The leaf stalk is weak and leaves often lie flat on the ground. Each spring new leaves appear which remain through the winter, often becoming reddish or mottled with bronze.

ASSETS The Hepatica is a charming native, especially desirable for its early bloom. It is comparatively easy to establish in a wild-flower garden and is available from all specialists.

CULTURE Plants can be lifted from the wild and in fact can be dug successfully even in flower, which is when the plants are most conspicuous. They need a part-shaded location and rich soil, preferably one containing some clay and quantities of organic matter such as peat moss or leafmold. The soil should be well drained but should never become bone-dry. Established clumps can be divided after flowering or in early fall. Colonies often increase by self-sowing.

SPECIES *Hepatica acutiloba.* Sharp-lobed Hepatica or Liverleaf. Native from Maine, Quebec and Ontario, south to Georgia and west to Missouri and Minnesota. (The two species named here are very similar and some botanists consider them mere variants of one species.) *H. acutiloba* has pointed leaf lobes and is found in slightly acid or neutral soil.

H. americana. Round-lobed Hepatica or Liverleaf. (Illustrated) Native from Nova Scotia and Manitoba, to northern Florida, west to Iowa and Missouri. This species is usually found in acid-soil areas.

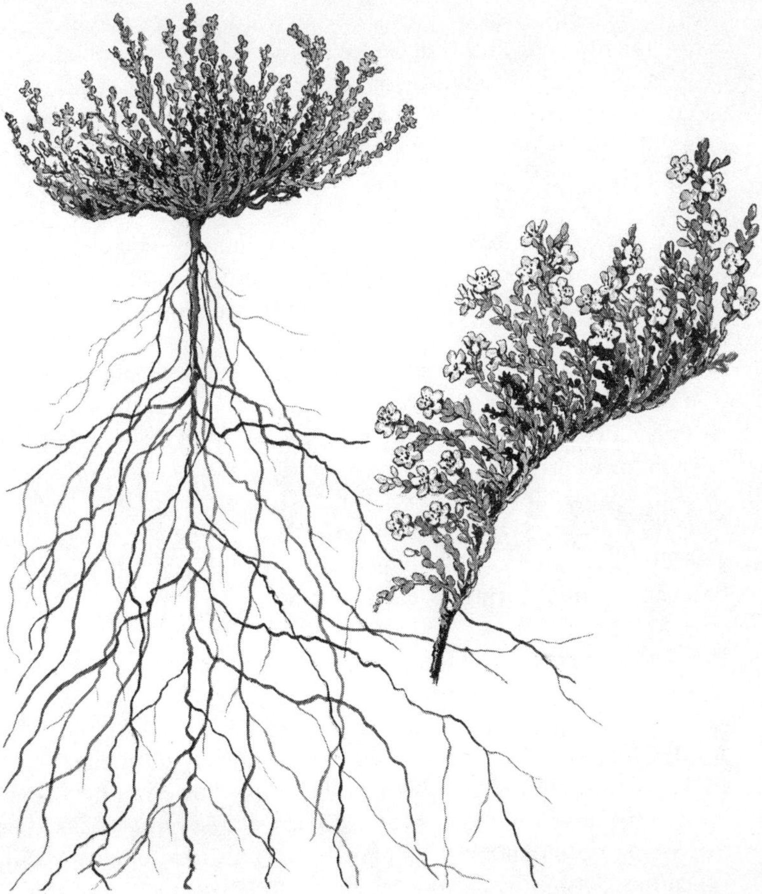

HUDSONIA

TYPE Evergreen sub-shrub
FAMILY *Cistaceae* (Siss-*tay*-see-ee)
GENUS *Hudsonia* (Hud-*soh*-nee-uh)
SPECIES *ericoides* (ehr-ik-*koy*-deez), *tomentosa* (toh-men-*toh*-suh)
ORIGIN See Species, below

This unique small shrub, which looks like a Heather but isn't even in the same family, grows like a weed in certain sandy-soil areas, usually near both salt and fresh water, but is impossible to establish in rich woodland soil. It is included here because it should be known, appreciated and used by those in sections where it is prevalent. Common names include Beach-heather and False-heather.

USES As ground cover and accent on dunes, slopes and hills or on any sunny, sandy ground. It will grow among rocks and stones and is suited to a dry, sunny rock garden.

HABIT OF GROWTH The plant forms a rounded silhouette of tufted branches, usually about 6 inches or so high. Its fine roots spread wide and deep, one of the reasons why these plants are difficult to transplant.

FLOWERS Five-petaled, golden-yellow flowers, barely one quarter of an inch across, crowded along the ends of the short branches, in mid-spring to early summer.

FOLIAGE Tiny, scalelike leaves which are evergreen, green or gray, depending on the species.

ASSETS A superb plant for seashore gardens and inland dunes in the Great Lakes region. The bright yellow flowers, which cover the plants, flash in the sun and appear for several weeks. Out of bloom, the mounds or hummocks of Hudsonia, often the only vegetation dotting sandy soil, are most attractive.

FAULTS Hudsonia is a native suited to special situations and can be enjoyed by most people only when they visit Pine barrens, the seashore and like areas.

CULTURE It grows in sandy, acid soil, low in humus and nutrients essential to the welfare of most plants. It is often found with scrub Pines and Bearberry, usually in full sun, but it will tolerate some shade. It is resistant to wind and shifting sand. Small plants can be transplanted after a rainfall (include as many of the fine roots as possible) into similar soil situations. They should be watered several times. Very interesting effects can be achieved around beach houses with Hudsonia and Bearberry. Plants can be set in informal groupings as ground covers, or individual plants can be used as accents by steps, rocks and walls. Plants of Hudsonia self-sow readily.

SPECIES *Hudsonia ericoides.* Poverty-weed. Beach-heather. Gold-heather. Cloth-of-gold. Native from Newfoundland and New Hampshire, south to Virginia and North Carolina. Height about 7 inches. Tiny, evergreen scalelike leaves, greener than those of the following species. Bright yellow flowers in late spring and early summer.

H. tomentosa. False-heather. Beach-heather. Woolly Gold-heather. (Illustrated) Native from New Brunswick down the Atlantic Coast to North Carolina, and west to Manitoba and Alberta and the sand dunes and hills of the Great Lakes Region and North Dakota. Differs from the preceding species in its gray, downy leaves.

INDIAN CUCUMBER-ROOT

TYPE Herbaceous perennial
FAMILY *Liliaceae* (Lil-ee-*ay*-see-ee)
GENUS *Medeola* (Med-*dee*-oh-luh)
SPECIES *virginiana* (vir-jin-ee-*ay*-nuh)
ORIGIN Nova Scotia to Ontario, south to Florida and west to Minnesota

Indian Cucumber-root gets its name from its edible tuberous root, which has the flavor of the Cucumber. It is closely related to the Trilliums.

USES A plant for deep woodlands and shade and damp, rich soil where it makes a suitable companion to Ferns, Merrybells, Bead-lily and Trilliums. It will thrive in bogs and swamps and in the moist soil along a shaded brook or pool.

HABIT OF GROWTH Stiffly upright plants, 1 to 2½ feet high, from long, white rhizomes. The plant makes a graceful pattern because of the leaf arrangement.

FLOWERS Small, nodding flowers, yellow-green, 1 inch long, in clusters at the top of the stem, in late spring. Though rather inconspicuous, each flower has three prominent red styles which contribute some distinction. Handsome, though, is the dark purple berry on a red stem which follows each flower and does not droop.

FOLIAGE The leaves, usually about six or seven, are in whorls, the first set appearing in the middle of the stem, to be followed by one or two more whorls, the top group forming only a few leaves from which the flowers and fruits appear. In spring the emerging stem and leaves are protected by a downy covering which later disappears.

ASSETS The interesting foliage arrangement and the showy, Cherry-like berries in fall.

FAULTS Not for open, dry woods. The flowers do not show up well.

CULTURE Grow the Indian Cucumber-root in shade and in acid, rich soil that remains moist. Add peat moss and/or leafmold liberally to the soil and for best effect, start with three or more roots.

IRIS

TYPE Herbaceous perennial
FAMILY *Iridaceae* (Eye-rid-*day*-see-ee)
GENUS *Iris* (*Eye*-riss)
SPECIES *cristata* (kris-*tay*-tuh), *fulva* (*ful*-vuh), *missouriensis* (miz-or-ee-*en*-siss), *verna* (*vern*-uh), *versicolor* (ver-sik-*kol*-or)
ORIGIN See Species, below

There are many native Iris species, and nearly every section of this country has its counterpart of the Blue Flag, *Iris versicolor,* itself widely distributed over the eastern part of North America.

USES All of the tall Flags are suited to naturalizing in moist meadows and at the edge of woodlands and at the edges of large pools and ponds, but many are too large for restricted plantings. They can be planted in the background of flower borders but are not really any better than available cultivated kinds like Japanese and Siberian Irises. The small species like *Iris cristata* and *verna* are so low-growing that they easily fit into a variety of situations: in rock gardens, as well as open woodlands, as foreground plantings for trees and shrubs, especially Azaleas and Rhododendrons, and on banks and slopes.

HABIT OF GROWTH Tall, heavy clumps with sword-shaped foliage, from 2 to 4 feet, or low, creeping plants, up to 4 or 6 inches in height; all from rhizomes.

FLOWERS Both individually and collectively, the Iris flower is one of the most beautiful in the plant kingdom. While the native Iris species lack the size, intricate color combinations and prominent beards of the garden varieties, they are nonetheless charming. The natives considered here are generally blue or blue-purple. The yellow tall Iris seen growing wild in low meadows, marshes and near ponds is *Iris pseudacorus,* an introduced species from Europe which has escaped from gardens and appears to be a native.

FOLIAGE Strong, sword-shaped leaves, 2 feet or more long, or short Grasslike leaves only a few inches high.

ASSETS Their beautiful flowers and handsome plant habit, and rugged constitution and adaptability (for the most part).

FAULTS Many of the natives, especially under good care, grow too fast and take up too much space unless one is trying to fill up large areas.

CULTURE These Iris species are quite easy to grow and can be transplanted at any stage of growth. Small sections can be removed from large clumps of the tall species and will have filled out nicely in a few seasons. Both the Crested Iris (*I. cristata*) and Vernal Iris (*I. verna*), which grow in loose sods or mats, should be planted with their rhizomes close to the surface. All are suitable for full sun. The two dwarf species will grow in light shade.

SPECIES *Iris cristata.* Crested Iris. (Illustrated) Native from Maryland south to Georgia and Alabama, and west to Missouri and Oklahoma, but hardy in the North. Its flowers are pale blue or lavender, their falls accented by an orange crest, and are carried on 6- to 8-inch stems. The grassy foliage may be 15 inches long but is usually less. The Crested Iris flowers in mid-spring, at the same time as many Azaleas. The slender rhizomes creep over the ground's surface and can cover an area, especially if the soil is moderately rich and well drained but capable of retaining moisture. It does quite well in drier-soil situations. Soil can be acid or neutral in reaction. There is a rather rare, white-flowered form of *Iris cristata.* A species from the Great Lakes Region, *I. lacustris,* is similar but has smaller flowers.

I. fulva. Copper Iris. One of the Louisiana species native to the Gulf Coast but also in Georgia and Illinois; hardy North. Its flowers are coppery red and several are carried on 2-foot stems in late spring. It likes good soil in full sun.

I. missouriensis. Native from South Dakota to British Columbia, Southern California and Arizona. One of the Blue Flags of the West with light blue flowers in late spring and early summer on 12- to 24-inch stems, the leaves lower than the stems so the flowers are not hidden.

I. verna. Vernal Iris. Pennsylvania to Georgia. A dwarf species with evergreen leaves like blades of Grass, its flowering stems barely 4 inches or less high, bearing sweet-scented lavender-blue flowers in mid-spring. This is a charming species which can be temperamental. It must have acid soil, rich in peat moss or leafmold. It grows in light shade.

I. versicolor. Blue Flag, native from Labrador to Manitoba, south to Virginia, west to Ohio, Wisconsin and Minnesota. Very pretty purple-blue flowers in late spring to early summer on 3- to 4-foot stems. Common to wetlands but adaptable to average soil that doesn't become bone-dry.

IRONWEED

TYPE Herbaceous perennial
FAMILY *Compositae* (Kom-*poz*-it-ee)
GENUS *Vernonia* (Ver-*noh*-nee-uh)
SPECIES *altissima* (al-*tiss*-im-uh), *noveboracensis* (noh-vee-bor-ass-*sen*-siss)
ORIGIN See Species, below

Resembling a tall, shaggy Aster, the Ironweed spreads its special glory over marshy meadows and thickets in late summer ana fall. Its companions are Joe-Pye Weed, Boneset, and various Goldenrods and Asters.

USES Ironweed is a majestic plant suitable for any extensive border planting of tall native plants, as accents at the edge of woods or in sunny, moist meadows or along semiwild drives or walks.

HABIT OF GROWTH A very tall, bushy plant which may be only 3 feet high but is more likely to reach 5 or even 9 to 10 feet.

FLOWERS They are royal purple, each flower about ½ inch across, with several making up a flat-topped cluster. Each stem has several flowering branches, borne in sufficient quantity to make an impressive flowering spectacle in late summer and fall.

FOLIAGE Alternate leaves, lance-shaped, up to 10 inches long, dark green in color.

ASSETS The Ironweed has a stately habit and richly colored fall flowers, and in a moist, sunny situation it is easy to grow.

FAULTS Not suitable for any small-scale planting.

CULTURE The Ironweed likes rich, moist soil such as is found in a marsh or wet meadow. It can be introduced to average garden conditions if water can be provided during dry periods. It tolerates acid or neutral soils. Transplant in spring or late fall.

SPECIES *Vernonia altissima.* Native from New York to Michigan and Nebraska, south to Florida and Louisiana. Under ideal conditions, plants may reach 10 feet, leaves, 1 foot.
V. noveboracensis. (Illustrated) Common Ironweed. Native from Massachusetts to North Carolina and Mississippi. Height, about 9 feet; leaves, about 10 inches long. (The specific name "noveboracensis" means pertaining to New York.)

JACK-IN-THE-PULPIT

TYPE Herbaceous perennial
FAMILY *Araceae* (Ar-*ray*-see-ee)
GENUS *Arisaema* (Ar-riss-*seem*-uh)
SPECIES *triphyllum* (trye-*fill*-um)
ORIGIN Nova Scotia to Florida, west to Minnesota, south to Kansas and Louisiana

Jack-in-the-pulpit is one of the best loved and most familiar wild flowers of spring. It is in the same family as Skunk-cabbage and the tropical Anthurium. Botanists are intrigued by the Jack-in-the-pulpit because, like the oyster, it may change its sex, at one stage producing male flowers, later changing to produce female flowers and seed. Jack-in-the-pulpit is also know as Indian Turnip because its tuberous roots were eaten by Indians. (The roots are not recommended for eating—apparently the Indians used them to test the bravery of young men who had to consume the acrid root without flinching, or they cooked them over and over to remove the bitterness which could burn the mouth.) Another species of this genus, Green-dragon or Dragon-root, *A. dracontium,* though less conspicuous than Jack-in-the-pulpit, is a good companion plant and is offered by some nurseries.

USES Even one plant makes a strong accent among Ferns and other spring wild flowers. Or plant colonies of Jack-in-the-pulpit in any shaded garden where they can give early interest before Hosta and other later plants have developed.

HABIT OF GROWTH Stately and upright, the height varying from 1 foot or less to 2 or 3 feet, depending somewhat on the richness of the soil.

FLOWERS The common name describes the flower exactly: the actual flowers are at the base of the club-shaped spadix, the "Jack," while his "pulpit" is the canopy over him which is green or striped in purple-brown, green or white. In late summer the flowers form a bright red berry cluster.

FOLIAGE Each leaf is composed of three leaflets, usually one or two to each flower stalk, carried on a long stem, often higher than the flower stem. There is a superficial resemblance to the foliage of Poison-ivy.

ASSETS Jack-in-the-pulpit is one of the most satisfying—and satisfactory—of all wild flowers to transplant to the home garden. It is always available from nurseries specializing in native plants;

on the other hand, gardeners should have no hesitation about digging it, even in flower, from the wild. It is one of the few native plants which offers a worth-while bonus—its bright red berries which follow the chartreuse and purple-brown spring effect. It tolerates boggy soil conditions.

FAULTS Not satisfactory in dry, poor soil or sun.

CULTURE Plant in spring, summer or fall, always in shade, and in humus-rich soil which may be neutral or acid. Moisture is a requirement; without it plants remain stunted. When collecting plants from the wild, dig deep enough to catch the fleshy root, and once planted, water well until established.

JACOB'S-LADDER

TYPE Herbaceous perennial
FAMILY *Polemoniaceae* (Pol-em-moh-nee-*ay*-see-ee)
GENUS *Polemonium* (Pol-em-*moh*-nee-um)
SPECIES *reptans* (*repp*-tanz), *richardsoni* (richard-*son*-eye)
ORIGIN See Species, below

The Jacob's-ladder or Greek-valerian cultivated in gardens is *Polemonium caeruleum,* a species introduced from Europe and a pretty companion for Tulips and Candytuft. The two native species included here are quite similar to the European species and are the ones most offered by wild-flower specialists, although there are several other species scattered through North America, especially in the Western mountains. All are equally desirable.

USES Grow along woodland paths, in rock gardens, or among naturalized Daffodils or around Rhododendrons and Azaleas.

HABIT OF GROWTH Low-growing, spreading clumps from 9 to 12 inches high.

FLOWERS Attractive bright blue-purple, bell-shaped flowers, in early to late spring, carried in terminal clusters above the foliage mounds.

FOLIAGE The leaves are alternate and divided pinnately to give a graceful, lacy effect. Apparently the common name, Jacob's-ladder, is an allusion to the ladderlike arrangement of the leaflets.

ASSETS Easy-to-establish natives, readily available from wild-flower specialists, and desirable for their pretty flowers and attractive foliage.

CULTURE Grow in sun or light shade in humus-rich soil that can retain some moisture. When planting, if the soil is lean and dries out quickly, mix a quantity of peat moss or leafmold into it. The clumps are easily divided after flowering. Established clumps tend to self-sow.

SPECIES *Polemonium reptans.* (Illustrated) Jacob's-ladder. Spring Polemonium. Native from New York to Alabama and Kansas. Light blue flowers, about ¾ inch long, in mid-spring. Height about 1 foot.

P. richardsoni. Native to arctic areas. Makes a mound of foliage about 9 inches high with the blue flowers produced on 2-foot stalks in late spring.

LADY-SLIPPER

TYPE Herbaceous perennial
FAMILY *Orchidaceae* (Or-kid-*day*-see-ee)
GENUS *Cypripedium* (Sip-rip-*peed*-ee-um)
SPECIES *acaule* (ak-*kaw*-lee), *calceolus* (kal-*kee*-oh-lus), *candidum* (*kan*-did-um), *reginae* (rej-*jye*-nee)
ORIGIN See Species, below

There is hardly a wild-flower enthusiast who is not acquainted with the Lady-slippers or Moccasin-flowers. And there is hardly a gardener who hasn't lifted a clump of Pink Moccasin-flower (illustrated), well surrounded by soil, to take home. Often the plant settles down in its new home and thrills the gardener by reappearing the next year and flowering. Often, though, it disappears and the gardener feels guilty that he had disturbed it in the first place. The admonition to leave such clumps alone seems less valid, though, as civilization pushes into the haunts of these and other native plants. Many are going to be destroyed anyway; and the careful gardener can be successful, if he can provide similar conditions to those enjoyed by these Orchids in nature.

USES All of the Lady-slippers are suitable for large and small woodland gardens, shady, naturalistic rock gardens and for the damp edges of bogs and marshes in light shade. They are attractive in the foreground of Azaleas, and flower at about the same time as *Iris cristata.*

HABIT OF GROWTH They have fleshy, "spidery" roots which spread horizontally from a small rhizome.

FLOWERS Unique and beautiful and impossible to confuse with any other kind of wild flower because of their shape, that of a slipper or sac, which is actually the lip of the flower. The surrounding petals and sepals are less conspicuous. The slipper color varies according to the species: it may be white, pink, rose, or yellow, and the sepals and petals a contrasting color. The Lady-slippers flower in spring and early summer.

FOLIAGE The foliage is attractive, differing according to the species. In the case of the Pink Lady-slipper or Moccasin-flower (*Cypripedium acaule*), there is a pair of basal leaves, while the Yellow Slipper Orchid (*C. calceolus pubescens*) presents leaves right up the flowering stalk.

ASSETS Their colorful flowers in a distinctive form.

FAULTS One could go on at length concerning the flaws of this genus, which are shared by most of our native Orchids when they are uprooted from their natural habitats and more or less forced into garden competition. (The tropical Orchids generally behave much better when introduced into home greenhouses.) The genus of Cypripedium in particular can be disappointing and fleeting in home wild-flower gardens, even when the gardener thinks he has provided proper conditions.

CULTURE Those who justifiably are challenged by these beautiful natives and can provide reasonably good conditions should order a few from wild-flower nurseries, or lift the kinds that are growing in the vicinity, especially if the area will soon be open to development. Specific recommendations for each species will be found at the end of this section, but in general Lady-slippers like cool conditions in summer and acid soil, heavy in humus-forming materials like leafmold and peat. After planting, a permanent mulch of Pine needles and leafmold should be maintained. Some require more moisture than others, specifically the Showy Lady-slipper (*C. reginae*), which needs even, constant moisture in its soil. The Pink Lady-slipper (*C. acaule*) thrives at the end of Long Island in bone-dry, very acid, sandy soil situations, but performs even better when observed nearby in more moist lowlands. Actually so little is understood about the requirements of most of our native Orchids that it is always a good practice when transplanting from the wild to dig a goodly amount of the surrounding soil with each clump. While many nurseries make a point of shipping some of our Orchids in "sods," this is obviously impractical for most species. When the bare roots are received, they are usually carefully wrapped and packed in sphagnum moss; it is a good idea to retain this material and place some of it around the roots when planting. The "eye" or buds above the rhizome should be just below the surface, about 1½ inches, the brittle fleshy roots spread out naturally. Seed of many Cypripedium species is available. Kathryn Taylor, an authority in the growing of native plants, recommends planting the seeds in soil from around mature plants which has been mixed with rotted Pine needles, leafmold and sand. Most species can be propagated by division of the tuberous roots, although once clumps are established in the wild-flower garden, most of us will be loath to disturb them for this purpose.

SPECIES *Cypripedium acaule.* Pink Lady-slipper. Pink Moccasin-flower. Native over a wide area in both dry, sandy situations, as well as near-bog and rocky slopes but always in very acid soil, from Newfoundland to Manitoba, south to Georgia and west to Minnesota, Nebraska, Kansas and Arkansas. This species is one of the first Orchids to flower in spring—around the middle of May on Long Island, later of course in colder areas. The single flower, on a stem 15 inches or less, is between two deep green, deep veined basal leaves. The lip, prominently covered with a network of red veins, may be deep pink or quite pale pink, the petals and sepals brownish. This beautiful species needs very acid soil, which may be dry or quite damp, but should be full of humus, preferably from rotting and decayed Pine needles and Oak leafmold, with some sand and peat moss completing the mixture. Renée du Pont Donaldson of Millerton, New York, recommends a yearly top-dressing of *rotted* cow manure over the clumps. Plants will be found growing in very light shade as well as deep shade. Every wild-flower nursery sells plants of the Pink Lady-slipper and it is worth growing even if it flowers for only a few seasons.

C. calceolus. Yellow Lady-slipper. This species has two varieties, the Large Yellow Lady-slipper, which is listed in catalogs as *C. pubescens,* and the similar but smaller Small Yellow Lady-slipper, *C. parviflorum.* The Large Lady-slipper inhabits rich, usually moist woodlands and swamp hummocks over much of North America from Nova Scotia and Ontario to Minnesota, Nebraska and Colorado and westward into Utah, Washington, Oregon and British Columbia. It also occurs in the south, including Georgia and Alabama, to Texas and New Mexico. Its slippers, of course, are a bright yellow, from 1 to 2 inches long, the sepals and petals twisted and colored purple-brown. The leafy flowering stem may be 1½ to 2 feet tall. Flowering time is late spring. Most authorities agree that this is the easiest Lady-slipper to move into the wild-flower garden. Give the roots a soil mixture of leafmold and/or peat moss and average garden soil near neutral or moderately acid. The soil should not become bone-dry. High shade is fine. Maintain a mulch over and around the clumps which may last for many years. The Small Yellow Lady-slipper is similar but fussier about its growing conditions, preferring more shade and neutral soils that are cool in summer and retain even moisture.

C. candidum. White Lady-slipper. Native in wet, grassy areas and woodland from Newfoundland to Manitoba, south to Georgia and Missouri. It has exquisite white slippers on stems from 6 to 12 inches high and should respond to a damp, humusy soil to which ground limestone has been added.

C. reginae. Showy Lady-slipper. (Still listed under its old name, *C. spectabile,* in some catalogs.) A very beautiful species, native in wet woods and bogs from Newfoundland to Ontario, Minnesota and Georgia. The white sepals and petals are more prominent in this species and make a spectacular contrast to the white slipper. The flower, sometimes in pairs, is on a leafy stout stem which is hairy and may reach a height of 2 or even 3 feet. Flowering time is early to midsummer, depending on how far north the plant is growing. The Showy Lady-slipper is the state flower of Minnesota. Again, this species needs a reasonably moist soil full of humus—peat moss, leafmold, compost—that is near neutral or only moderately acid. The plant's crown should be level with, not below, the surface of the soil. It is a late starter in spring, so mark its location well so the shoots just below the surface aren't injured. It should have high shade and good air drainage.

NATIVE PLANTS FOR ROCK GARDENS

The Hyssop-leaved Eupatorium in the foreground of this boulder needs sun and dry, sandy soil. In summer its fuzzy white flower heads make an effective cover in areas where many other plants won't grow. The yellow-flowered Golden-star, below, does best in some shade and is handsome among rocks and in retaining walls.

Anyone who is fortunate enough to have the makings of a rock garden nearby or to inherit a series of rock outcroppings can utilize many native plants. This large rock garden is being built by the John Knippenbergs in New Jersey and so far includes Bluets, Ferns and Trailing-arbutus.

This woodland rock garden, part of a conservation project of the Pines Lake Garden Club with the local school, includes Ferns, Jack-in-the-pulpit, Solomon's-seal, and Hepatica.

GARDEN POOLS OF ALL KINDS OFFER SETTINGS FOR WILD FLOWERS

Clumps of Foamflower thrive in rich soil at the edge of this garden pool in partial shade.

This pond at the Berkshire Garden Center in Massachusetts is planted with native water plants as well as those preferring the moist soil found at its edges.

THE PINK LADY-SLIPPER may last for years in a wild-flower garden if acid soil is provided. This clump is growing in near-boggy conditions but the plants also thrive in quite dry, sandy soil that is acid.

LILY

TYPE Herbaceous perennial
FAMILY *Liliaceae* (Lil-ee-*ay*-see-ee)
GENUS *Lilium* (*Lill*-ee-um)
SPECIES *canadense* (kan-ad-*den*-seh), *grayi* (*gray*-eye), *michauxi* (mee-*show*-eye), *michiganense* (michigan-*en*-seh), *pardalinum* (par-dal-*lye*-num), *philadelphicum* (fil-ad-*delf*-ik-um), *superbum* (soo-*perb*-um), *washingtonianum* (wash-ing-ton-ee-*ay*-num)
ORIGIN See Species, below

Most of our native Lilies bloom in early summer, a between-time when their bright, spectacular flowers are especially appreciated. Don't confuse these with non-native colonies of the Day-lily (*Hemerocallis*)—really escapes from nearby former gardens—which flower at about the same time. The Day-lilies, with their Lilylike flowers, are often called Tiger-lilies, a harmless enough error, except that confusion is compounded because the name "Tiger" is also applied to a true Lily, also an escape from gardens, which flowers around the same time. This Tiger Lily (*Lilium tigrinum*) is native to Japan and China and has long been cultivated in American gardens. It has orange flowers, heavily spotted, and always has a cache of little black bulblets in its leaf axils. These little bulbs, of course, are one reason why it spreads from cultivation so easily. The Tiger Lily, which thrives under a variety of conditions, sun or light shade, varying soils, is worth naturalizing among truly native species.

USES The Lilies included here are fine for establishing in any open woodland or at the edge of a woodland. Most are so tall that they need the companionship of shrubs and small trees. Several are tolerant of wet soil and will grow at the edges of ponds or pools or in wet meadows. They associate especially well with broad-leaved evergreens like the Rhododendron.

HABIT OF GROWTH Tall, from 3 to even 8 feet; sometimes spreading into colonies from their scaly bulbs, but occasionally from rhizomes.

FLOWERS The flowers of a Lily may be bell-, trumpet- or cup-shaped, with the petals reflexed or just slightly curved as in the case of *Lilium canadense,* the Canada Lily. They can be nodding or drooping or up-faced, as is the Wood Lily (*L. philadelphicum*). Usually more than one flower is produced. Some of the natives have slight fragrance.

FOLIAGE Whorls of leaves at intervals up the stems, the leaves being lance-shaped like those of a Willow, several inches long.

ASSETS The Lily as a whole—its bulb, its architecturally perfect leaf arrangement and its beautiful flowers—adds up to a majestic plant, and our native species, many of which remain untouched by the hybridizer, compare favorably with both non-native species as well as the lovely garden hybrids. Most of these natives are not difficult to grow, and their flowers, especially long-lasting, appear after the lavish spring showing made by native plants but before most of the summer plants have started to flower.

FAULTS Most of these Lilies need rich, even moist soil that is acid, and some shade. The bulbs are often destroyed by rodents. A few are so tall that it takes ingenuity to fit them into an intimate, small-scale naturalized garden.

CULTURE Many Lilies form roots along their stems, which is why such deep planting of the bulbs is recommended for certain species. Propagation of all kinds of Lilies includes seeds (sown in fall for spring germination), scales (leaflike sections of the bulb which are carefully peeled off and planted about ½ inch deep in sand and peat moss), bulb separation in the fall, and by bulbils which may form above or below the ground on the stem, as in the case of the Tiger Lily. Wild-flower nurseries offer bulbs for most of our natives which should be planted in the fall. Soil requirements for each species will be found under the descriptions of the species.

SPECIES *Lilium canadense.* Meadow Lily. Canada Lily. Native from Nova Scotia to Georgia and Alabama, west to Minnesota and Nebraska. Bright yellow, orange or red flowers spotted with brown, drooping and several in number, on long stems at the end of a stalk which may be from 3 to 5 feet or more tall. The lance-shaped leaves are in graceful whorls up the stem. Flowering time is early to midsummer. This Lily will grow in sun or light shade in average soil, neutral or acid, which retains moisture. Bulbs should be planted in the fall, about 6 to 8 inches deep. Its height makes it a good subject for planting in open woodlands among or near informal shrubs or small trees.

L. grayi. Gray's Lily. Roan Lily. Virginia, North Carolina, and Tennessee. This species may bear one or more, usually up to six, drooping flowers at the end of its 2- to 5-foot stem in early summer. The flowers, without reflexed petals, are bell-shaped and red with accents of yellow at their base. Each flower is heavily spotted with brown. It will grow in sun or part shade but must have a soil which retains moisture. Plant the bulbs about 4 inches deep, mixing a quantity of acid leafmold and/or peat moss in the hole, and if the soil is heavy clay, add some sand as well. This is another attractive species for woodlands, informal shrub borders and around pools.

L. michauxi. Carolina Lily. This species, formerly known as *L. carolinianum,* is another Southern species, hardy well above its native areas in Virginia to Florida and Louisiana. Its fragrant, drooping flowers are orange-red with purple spots and appear on 3- to 4-foot stems in early summer. Plant the bulbs 4 to 5 inches deep in soil rich in acid leafmold which will retain moisture, in sun or part shade.

L. michiganense. Midland Lily. Native from Michigan to Minnesota and Missouri, it is similar to *L. canadense* and is found in moist prairie land. The flowers, bright orange and spotted with brown, are much more reflexed in their petal formation than are those of the Meadow Lily. Usually one or two flowers are produced at the ends of the stems, 2 or 3 feet high, in early summer. Of interest is the fact that the small, scaly bulb of this Lily forms horizontal rhizomes, more like those of an Iris than a Lily. If one is planted, about 4 inches deep, it will soon spread to form a colony several feet in all directions. These rhizomes will form their own bulbs so that the original plant can be multiplied many times. Grow in sun or light shade in rich soil that retains moisture and which is full of organic matter like leafmold, peat moss or compost.

L. pardalinum. Leopard Lily. Oregon to California in the coastal ranges. A Western species with lovely orange-red flowers in early summer on tall stems which can reach 10 feet. As many as thirty-two flowers have been recorded on plants in the wild. This species needs cool woodland conditions and rich, acid soil that retains moisture. Plant the bulbs about 5 inches deep. This is another species that spreads by horizontal rhizomes.

L. philadelphicum. Wood Lily. Orangecup Lily. Native from Maine to British Columbia, south to North Carolina, Nebraska and New Mexico. One, rarely more than two, cup-shaped, up-facing flowers, orange or yellow, the inner petals speckled with brown, in early summer. This species grows only 3 feet tall, making it useful for small-area naturalized plantings, but unfortunately it has a reputation for being difficult. It needs acid soil full of humus, and is often found in rather dry situations, in part shade. It is especially lovely among Ferns and other leafy native plants like Alum-root and Foam-flower. If mice are a problem, this is one Lily that might be worth the trouble of planting in a narrow-mesh wire basket. Plant the bulbs 5 inches deep.

L. superbum. American Turk's-cap Lily. (Illustrated) Native from New Brunswick to Minnesota, south to Florida and Alabama. This late summer-flowering Lily is truly majestic, each 6- to 8-foot stem carrying about seven flowers at the top, although specimens have been recorded with as many as forty flowers. Plant the bulbs 4 to 5 inches deep in soil enriched by quantities of leafmold, peat moss or compost. It needs acid soil and is often found in bogs and swamps, but it will tolerate somewhat drier soil. Its height is its only drawback; plant it among tall shrubs.

L. washingtonianum. Shasta Lily. Mount Hood Lily. Oregon and California. Western species with small, white trumpet-shaped flowers which turn purple as they age, on 5-foot stems in late spring or early summer. Although it inhabits rich, cool Fir forests in the West, this Lily seems to do quite well elsewhere in humusy soil and some shade. Plant the bulbs about 5 to 6 inches deep.

LOBELIA

TYPE Herbaceous perennial
FAMILY *Lobeliaceae* (Loh-beel-ee-*ay*-see-ee)
GENUS *Lobelia* (Loh-*beel*-ee-uh)
SPECIES *cardinalis* (kard-in-*nay*-liss), *siphilitica* (sif-il-*lit*-ik-uh)
ORIGIN See Species, below

Nature lovers cherish particularly the memory of a colony of Cardinal-flowers (*Lobelia cardinalis*) in full bloom in midsummer. While stands of this beautiful plant, whose scarlet flowers do mirror the bright feathers of the cardinal, are not as plentiful in the wild as they used to be, it is still possible to find clumps, even colonies, and special trips into the country for this purpose can be rewarding. Fortunately the Cardinal-flower is not difficult to cultivate. Less spectacular but lovely in its own way is the Great Blue Lobelia, *L. siphilitica,* which is illustrated.

USES Both species can be grown as accents in sun-streaked woodlands or meadows, and are especially effective in waterside locations such as along brooks and streams or beside pools, either natural or man-made. The Cardinal-flower can be spectacular in a large, naturalistic rock garden or at the base of a boulder or in the crevices of rock outcroppings. Both species make excellent companions to Ferns. They can be grown in semi-shady parts of the perennial garden.

HABIT OF GROWTH The Cardinal-flower forms a tall clump, from 2 to 4 feet, depending on soil moisture and fertility; the Great Blue Lobelia is also tall, but more bushy in habit, carrying many side shoots which also produce flowers.

FLOWERS The Cardinal-flower has 1½-inch, long-tubed flowers, bright scarlet on a long spike or raceme, in early summer, often into the fall. Howard S. Bradley of Hamilton, New York, has dedicated himself to the project of collecting white-flowered forms of many of our native plants. He is responsible for adding the rare white form of the Cardinal-flower to the lists of many wild-flower nurseries. The Great Blue Lobelia has 1-inch-long, tubular violet-blue flowers, the petals of the lower lip being prominently marked with white. The tube is striped blue and white on its bottom side. The flowering period is long, usually from late summer into fall. There is also a white form of this species.

FOLIAGE Alternate lance-shaped leaves, about 5 inches long, sometimes with a bronze cast as in the leaves of the Cardinal-flower. The Great Blue Lobelia produces its bright green leaves right up to the end of the flowering stems.

ASSETS The Cardinal-flower in full bloom is one of the glories of the late wild-flower garden, and its companion, the blue-flowered Lobelia, has great charm. Both, either en masse or as accent clumps, are suited to a variety of situations and climates. Both are comparatively easy to establish and are sold by most wild-flower nurseries. Their flowering season lasts for many weeks.

FAULTS By late summer, the Great Blue Lobelia may form an immense bush, with so many side branches that the plant falls over and, in a small wild-flower garden, can smother nearby plants. The solution, of course, is to give clumps plenty of room and to keep delicate plants out of their environs. The Cardinal-flower may die out for no apparent reason.

CULTURE Moist, humus-rich and acid soil is preferred by both species, although they will persist fairly well in average soils if water is applied during droughts. If the soil is moist, the Cardinal-flower tolerates full sun, but both species seem to prefer part shade. The Great Blue Lobelia will grow in quite dense shade. Both species are free with their seeds and may self-sow freely. Seeds can be sown in spring or fall. Plants can also be divided in spring. It is probably best to remove the fading spikes of the Cardinal-flower unless a seed harvest is desired.

SPECIES *Lobelia cardinalis.* Cardinal-flower. Cardinal Lobelia. Native from New Brunswick, west to Ontario, Kansas and Colorado, south to Florida and Texas. Flowering stalks may reach 6 feet in height when soil is especially rich and moist.
L. siphilitica. Great Blue Lobelia. Found from Maine and Ontario to South Dakota and Colorado, south to Georgia, Louisiana and Kansas. Although less familiar than the preceding species, it should be sought out and will be appreciated for its blue flowers over a long period.

LUPINE

TYPE Herbaceous perennial
FAMILY *Leguminosae* (Leg-yew-min-*noh*-see)
GENUS *Lupinus* (Lew-*pye*-nus)
SPECIES *perennis* (per-*ren*-niss)
ORIGIN Eastern half of United States

While this wild blue Lupine is widely distributed, it is found only in sandy, dry and acid soil. It is especially common in Pine-barren regions, its spikes of bright blue flowers making a rich color display in mid- to late spring. It has been called the Sun-dial Lupine. (Also of interest are the Bluebonnets of Texas, annual Lupines which can be grown out of their state when one can procure the seeds. Two Bluebonnet species are *Lupinus subcarnosus* and *texensis*. California has several Lupines, including the handsome Tree Lupine, *L. arboreus,* suitable only for mild-climate regions. The Pacific Northwest has *L. polyphyllus,* a hardy perennial Lupine from which such beautiful garden hybrids as the Russell Lupines are derived.)

USES For naturalizing in any open dry woodland or sandy-soil areas, especially on banks and slopes and along paths and roadways.

HABIT OF GROWTH This Lupine can form clumps with many flower spikes almost as tall and handsome as those of our more refined garden Lupines, but more often the plants are sprawling, not growing much higher than 15 to 18 inches. The roots of mature plants are woody and slender, sometimes going down as deep as a foot and spreading as much horizontally.

FLOWERS Very pretty blue, blue-purple or pinkish flowers in tall spikes, sometimes marked with white, and typical of the Pea family, in late spring.

FOLIAGE Compound leaves, palmately divided into small leaflets just like our garden hybrids.

ASSETS Their lovely blue flowers, which last well when arranged indoors, and the plant's tolerance of dry, sandy soil.

FAULTS Not for rich soil. Mature plants can not be dug successfully.

CULTURE Anyone who lives in regions where this Lupine abounds naturally can collect little seedlings in late spring and plant them in acid soil that is gravelly and sandy. A second method is to collect seed just as soon as ripe in early summer and sow immediately in a mixture of sand and peat moss or leafmold. Germination is erratic and most of the seedlings probably won't appear until the following spring. Although flowering clumps of this Lupine often look easy to dig, and nearly everyone living near them has tried, it's an impossible task and failure is bound to result.

MARSH-MARIGOLD

TYPE Herbaceous perennial
FAMILY *Ranunculaceae* (Ran-nun-kew-*lay*-see-ee)
GENUS *Caltha* (*Kalth*-uh)
SPECIES *palustris* (pal-*lust*-triss)
ORIGIN-Newfoundland to Saskatchewan, south to South Carolina and Nebraska

The only resemblance this native bears to the Marigold at all is the color of its golden flowers. Another common name is Cowslip or American Cowslip. In England, where this same species is found, it is known as King-cup and May-blob, the latter being especially descriptive. It is in the Buttercup family.

USES Colonies of the Marsh-marigold will brighten any wet-soil situation such as may exist around pools, man-made or natural, and along brooks where it is usually found in nature. It will, of course, thrive in bogs and marshes.

HABIT OF GROWTH Rounded clumps of succulent foliage about 1 foot high.

FLOWERS The bright yellow flowers, about 2 inches across, are carried above the leaves on 1- to 2-foot hollow stems, in early spring. There is a rare, double-flowered form, listed as *monstrosa-pleno*.

FOLIAGE Rounded or heart-shaped glossy leaves, which vary from 2 to 7 inches in width. The foliage usually dies down in summer when the plant goes dormant.

ASSETS The Marsh-marigold is one of our most handsome natives of early spring, its combination of bright yellow flowers and glossy leaves being an especially happy one. It is easy to grow if moist conditions can be provided and is available from all wild-flower specialists.

FAULTS In the home garden, plants of Marsh-marigold may become "lost," as they go dormant in summer.

CULTURE Extra moisture is essential to the Marsh-marigold in spring, when it is often found actually growing in water. Later, as spring floods subside, it is tolerant of average soil conditions, a factor to be considered when introducing it into the home wild-flower garden and one that makes it comparatively easy to grow. Its soil should be rich in humus; its exposure sun or light shade. Plants are easily divided in early spring or immediately after flowering by separating the fibrous-rooted clumps.

MAY-APPLE

TYPE Herbaceous perennial
FAMILY *Berberidaceae* (Ber-ber-id-*day*-see-ee)
GENUS *Podophyllum* (Pod-oh-*fill*-um)
SPECIES *peltatum* (pel-*tay*-tum)
ORIGIN Quebec to Minnesota, south to Florida and Texas

Vast colonies of May-apple (or Mandrake) are so conspicuous in the woods in spring that this native is known by those with only a casual interest in wild flowers. The common name May-apple probably refers to the time the flowers appear and to the apple-shaped fruit, which is conspicuous in mid- to late summer. The term "Mandrake" rightly belongs to a European plant but was probably passed on to May-apple because its roots and foliage are very poisonous. The fruit, however, is edible and was once used to make a preserve.

USES A handsome ground cover and accent for shade on banks, along fences and walls, walks and drives, wherever far-spreading patches of greenery are desirable.

HABIT OF GROWTH From the root, a rhizome, rises a thick stem about 12 inches high, which bears from one to three leaves.

FLOWERS The fragrant white flowers, which appear singly from the axil between two leaves, are from 1½ to 2 inches across and in both the bud and flowering stage are attractive, but unfortunately they are usually hidden by the leaves. The edible fruit which follows is yellow, about 2 inches long and oval. It is packed with seeds.

FOLIAGE The segmented leaves of May-apple, deeply lobed and toothed, are its most intriguing part, and very handsome they are. Each leaf, as it emerges from ground in early spring, is tightly furled downward like a half-open umbrella. Eventually it fully opens, making a canopy over the ground. The color of the leaves is interesting too, the upper surfaces being a dark green, the under surface, a pleasing light green. These effects are especially apparent when plants are observed from the bottom of a slope.

ASSETS The May-apple is tolerant of a variety of soils and exposures and is fast-spreading. It can be naturalized on steep slopes. It is a good plant for roadsides, thickets and waste areas too dense for less weedy natives. Either in big colonies or isolated plantings, the May-apple is a distinctive native plant.

FAULTS May-apple can become a nuisance, soon overrunning more delicate plants. In a small wild-flower area, it should be omitted unless the gardener is willing to keep a wary eye on its spread. This spreading fault can be used to advantage by those who have inherited large open woodlands and wish to establish a quick-growing, handsome ground cover.

CULTURE Ideally May-apple likes moist, humus-rich soil in light shade but grows quite well in conditions far from this ideal. The soil may be neutral or acid. Once introduced, the plants will propagate themselves by seed or by their spreading rhizomes. Plants can be lifted from the wild easily at any time, although early spring is the best time. May-apple is sold by all wild-flower specialists.

MEADOW-BEAUTY

TYPE Herbaceous perennial
FAMILY *Melastomaceae* (Mel-last-tow-*may*-see-ee)
GENUS *Rhexia* (*Rex*-ee-uh)
SPECIES *mariana* (mar-ee-*ay*-nuh), *virginica* (vir-*jin*-ik-ah)
ORIGIN See Species, below

A beauty, indeed, is this well-named summer-flowering plant. As more and more moist meadows and grassy bogs disappear, so will such gems as Meadow-beauty, unless the home gardener is willing to go to a little trouble to provide the special conditions such plants require.

USES Meadow-beauty grows naturally in wet soil so is a perfect subject for a bog garden, either natural or man-made, or around pools. It will also grow in any rich soil that remains reasonably moist, so it is suitable for the naturalistic rockery and along walls.

HABIT OF GROWTH Bushy plants, with winged stems up to 1 or 2 feet, from small swollen roots, really tubers.

FLOWERS Rosy pink or bright magenta flowers in clusters at the tops of the stalks. Each flower, about 1 inch across, has four petals accented by bright yellow anthers and, alas, lasting barely a day. Flowering time is from early summer into fall, the greatest display being at midsummer.

FOLIAGE Opposite leaves, somewhat hairy, which clasp the stems.

ASSETS The lovely, fragile-appearing bright flowers.

FAULTS Meadow-beauty must have acid soil.

CULTURE Moist, acid soil of peat and sand and full sun are preferred, but Meadow-beauty does quite well in adaptations of this so long as the soil is acid. Plenty of peat moss should be mixed with the home-garden soil. The tuberous roots of *R. virginica* can be separated in spring or fall. Seeds of Meadow-beauty are available and should be sown in the fall in a mixture of sand and peat moss for germination in the spring.

SPECIES *Rhexia mariana.* Maryland Meadow-beauty. Native in moist sand-and-peat soils from Long Island to Northern Florida and Texas. Height about 2 feet. Flowers pale pink with bright yellow stamens in summer. This species does not have swollen roots and is just about impossible to dig from the wild when in full bloom, which is about the only time the plant collector sees it. Seeds, when available, provide the best means of growing this species.

R. virginica. Meadow-beauty. Deer-grass. (Illustrated) Native from Nova Scotia to Ontario and Wisconsin, south to Georgia and Oklahoma. Height about 3 feet, its stems definitely "winged" or angled. Flowers are rosy purple. Fairly easy to grow because of its tuberous roots. Plants and seeds of this species are offered by several nurseries.

MEADOW-RUE

TYPE Herbaceous perennial
FAMILY *Ranunculaceae* (Ran-nun-kew-*lay*-see-ee)
GENUS *Thalictrum* (Thal-*lik*-trum)
SPECIES *dasycarpum* (das-ik-*karp*-um), *dioicum* (dye-oh-*ik*-um), *polygamum* (pol-*lig*-am-um), *venulosum* (ven-yew-*loh*-sum)
ORIGIN See Species, below

The engaging Meadow-rue, in some of its European and Asiatic species, is often included in rock gardens and perennial borders. The native species listed here have the same airy foliage as the cultivated Meadow-rues and are generally no more notable for their flowers than are the cultivated kinds.

USES Meadow-rues are effective in colonies or as ground covers in open woodland or in full sun, in naturalistic rock gardens or against stone walls.

HABIT OF GROWTH Mounds of foliage from which rise branching and leafy stalks with small, fuzzy flowers.

FLOWERS Each small flower, displayed in clusters or panicles, has no petals but several stamens, which give it its fuzzy appearance. The flowers may be one-sexed or complete, with organs of both sexes represented in one flower. The colors of the flowers are not especially bright and are usually green, purplish, white or yellow. Blooming time varies with the species, from spring to summer.

FOLIAGE The foliage of the Meadow-rue often is compared to that of the Maidenhair Fern and Columbine. The compound leaves, blue-green, are made up of a series of three leaflets, each leaflet being deeply lobed.

ASSETS There is no question but that the foliage of these native Meadow-rues is their most worth-while characteristic. They are also easy to grow and transplant and are tolerant of varying soil conditions.

FAULTS Not all species have very conspicuous flowers.

CULTURE The Meadow-rues generally prefer a rich, humusy soil on the moist side, although they are often found in average soil. They thrive in sun or light shade. Most wild-flower nurseries offer plants which may be set out in spring or fall. Established clumps can be divided in early spring or after flowering. Wild plants can be transplanted at any time if the soil is moist and they are watered after planting.

SPECIES *Thalictrum dasycarpum.* Purple Meadow-rue. Native from Ontario and Manitoba, south to Ohio, Louisiana, Texas and Arizona. Its stems are purple, from 3 to 6 feet high, and bear white and purple flowers, showy in clusters in summer. A tall species for rich, moist soil, best in sun or light shade.

T. dioicum. Early Meadow-rue. Native from Quebec to Manitoba, south to Alabama, Missouri and North Dakota. Spring-flowering, its green-yellow flowers drooping in panicles, on stalks up to 2 feet. Its compound leaves, long-stemmed, may reach or rise above the flowers. An attractive species for light woodlands and effective with Columbine, Ferns and native Lilies.

T. polygamum. Tall or Fall Meadow-rue. (Illustrated) Native from Newfoundland to Ontario, south to Florida, and Ohio. Another tall species, from 3 to 5 feet or more, which produces its pure white flowers in summer or early fall. The flowers are especially effective against the airy foliage. At its best in rich, moist (even wet) soil.

T. venulosum. Native from Manitoba to Utah. This western species has attractive mounds of blue-green foliage with yellow- or bronze-tinged flowers which are not especially attractive.

MERRYBELLS

TYPE Herbaceous perennial
FAMILY *Lilicaceae* (Lil-ee-*ay*-see-ee)
GENUS *Uvularia* (Yew-vew-*lay*-ree-uh)
SPECIES *grandiflora* (gran-dif-*floh*-ruh), *perfoliata* (per-foh-lee-*ay*-tuh), *sessilifolia* (sess-il-if-*foh*-lee-uh)
ORIGIN See Species, below

Waiting to be discovered are the lovely, drooping flowers of Merrybells, also known as Bellwort, which too often seem to blend with the foliage and remain hidden. These flowers, much larger versions of those of the Solomon's-seal and Twisted-stalk, have tremendous appeal, once the observer slows down enough to appreciate them.

USES For shade or woodland, under trees or on shaded slopes or banks, or in naturalistic rock gardens and in the company of Rhododendrons and other shrubs. They combine well with Ferns and Trilliums.

HABIT OF GROWTH Upright, angularly forked or branching stems which are leafy. The root is a creeping rhizome which may produce several flowering stems, usually about 1½ feet high.

FLOWERS They are long, slender bells, greenish yellow or bright yellow, and hang from the tips of the branches and leaf axils in spring.

FOLIAGE Merrybells has attractive foliage. The leaves are alternate, pale green, lance-shaped and from 1 to 3 inches long, and encircle or clasp the stems.

ASSETS Graceful plants for the woodland and other shaded areas. They are easy to grow.

CULTURE The Merrybells will spread to form open colonies if they are planted in humusy soil (add peat moss or leafmold) that remains reasonably moist and never becomes bone-dry. As with many of our spring-flowering natives, some sunshine in spring is appreciated, but during summer they need shade to keep their roots cool. The plants are increased by cutting the creeping rhizomes apart in early spring or fall.

SPECIES *Uvularia grandiflora.* Great Merrybells. Native from Quebec to Ontario, Minnesota and Kansas, south to Georgia and Oklahoma. Lemon-yellow bells, about 1½ inches long, in late spring. The leaves clasp the stems, which may reach 1½ feet. A most attractive native, easily established in rich, slightly acid or neutral soil and part shade.

U. perfoliata. Smooth Merrybells. Quebec to Florida and Mississippi. Slightly smaller flowers which are pale yellow, about 1¼ inches long, and hang from 1½-inch stems. Stems and leaves are smooth, the leaves, blue-green in color. Spring-blooming.

U. sessilifolia. Strawbell. (Illustrated) Wild-oats. Native from New Brunswick and Ontario to Minnesota, south to Georgia and Arkansas. Its greenish flowers seem to blend with the pale leaves which are often furled as the flowers appear. Nevertheless it is most attractive and easily established in acid, rich woodland soil. Its height is about 1 foot, the bells about 1¼ inches long. Spring-blooming.

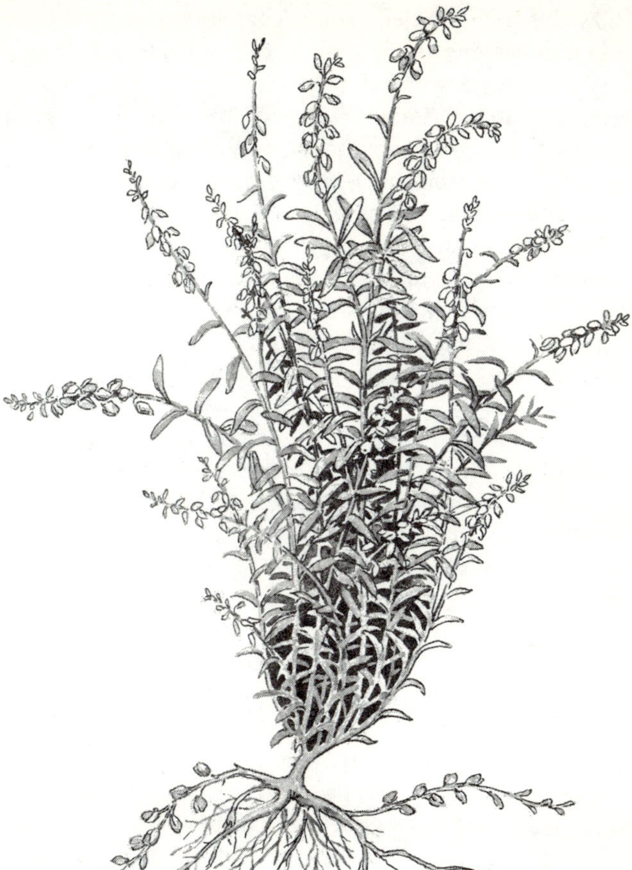

MILKWORT

TYPE Herbaceous biennial or perennial
FAMILY *Polygalaceae* (Pol-lig-uh-*lay*-see-ee)
GENUS *Polygala* (Pol-*lig*-uh-luh)
SPECIES *paucifolia* (paw-sif-*foh*-lee-uh), *polygama* (pol-*lig*-am-uh), *senega* (*sen*-eg-uh)
ORIGIN See Species, below

The Milkworts are a varied group, containing annuals, biennials and perennials among the native species. (There are several interesting moist-meadow inhabitants among the Milkworts which botanists prefer to call Candy-root or Candy-weed because of the Wintergreen scent and flavor of the roots, noticeable when they are crushed. Such species include *Polygala sanguinea,* an annual, *P. lutea,* a biennial and *P. cruciata,* Marsh Milkwort, an annual. All are attractive plants which would be worth introducing to moist, sunny locations in the wild-flower garden if seeds were available.) The Fringed Milkwort (*P. paucifolia*) is one of our choicest woodland natives and bears little resemblance to other Milkworts.

USES The Fringed Milkwort should be planted in colonies or groups in the woodland, especially against low boulders or in bays along a pathway. *P. senega,* the Seneca Snakeroot, is much taller and less refined but makes an effective display in colonies in open woodlands. The Racemed Milkwort (*P. polygama*) is similar to the preceding and is effective in groups in light shade or sun.

HABIT OF GROWTH The Fringed Milkwort is semi-evergreen, low-growing and sprawling, rarely exceeding 6 inches in height. Both the Racemed Milkwort and Seneca Snakeroot are bushy, upright plants with many stems, about a foot high. All three produce underground flowers which do not develop color but may set seed by means of self-fertilization.

FLOWERS The Fringed Milkwort has by far the most showy flowers with rose-purple sepals (like "wings") and petals, nearly 1 inch long in late spring and early summer. In fact other names for this species, which imaginatively describe the flowers, are Bird-on-the-wing and Gaywings. The smaller flowers of the Seneca Snakeroot and Racemed Milkwort are produced in terminal racemes in early summer.

FOLIAGE The persistent or semi-evergreen foliage of the Fringed Milkwort, though produced in sparse clusters toward the ends of the stems, is attractive, especially when a number of plants dot a woodland carpeted with Pine needles. Each leaf is about 1 to 1½ inches long. The foliage of both other species is carried in greater abundance, and although small and narrow, covers all the stems from the crown to the flowering tip.

ASSETS The Fringed Milkwort, with its graceful yet odd-shaped flowers, can be lovely beyond description; both the rose-colored flower spikes of Racemed Milkwort and the less showy white spikes of the Seneca Milkwort possess a pert charm, and the growth habits of the plants are satisfactory. The last two species are suitable for lean, sandy soil.

FAULTS The Fringed Milkwort is difficult to move from the wild because of its straggling underground stems and sparse roots.

CULTURE Shade and cool, moist soil rich in acid humus is the preference of Fringed Milkwort. Collected plants often languish. It is better to take stem cuttings early in summer after flowering is over, insert in a flat of peat moss and sand and keep shaded and moist. The cuttings should be rooted by fall and can be set out then or the following spring. Seeds, when available, can give fair results. The other two Polygala species, by comparison, are easy to grow, thriving in well-drained, even dry, acid, sandy soil in full sun or light shade. With care, plants can be lifted in full bloom and re-established if carefully watered at first.

SPECIES *Polygala paucifolia.* Fringed Polygala. Fringed Milkwort. Flowering-wintergreen. Gaywings. Native from New Brunswick to Saskatchewan, south to Georgia, Illinois and Minnesota. Evergreen plants, 3 to 6 inches high, with showy rose-magenta flowers somewhat resembling some Orchids.

P. polygama. Racemed Milkwort or Polygala. (Illustrated) Native from Quebec to Minnesota, south to northern Florida and Texas, in dry sandy fields. Small, rose-purple flowers in racemes about 1 foot high, in late spring and summer.

P. senega. Seneca Snakeroot. Similar to the preceding but with racemes of white flowers earlier in the season.

MONKEY-FLOWER

TYPE Herbaceous perennial
FAMILY *Scrophulariaceae* (Skrof-yew-lay-ree-*ay*-see-ee)
GENUS *Mimulus* (*Mim*-yew-lus)
SPECIES *ringens* (*rin*-jenz)
ORIGIN Nova Scotia, Manitoba to Virginia and Texas

It's quite evident from the shape of the flowers that this Monkey-flower and a similar species, *Mimulus alatus*, are in the Snapdragon family. The species above is the one most offered by wild-flower nurseries, but there are many other species native to North America, some with attractive flowers.

USES In naturalistic borders in light shade or sun and in moist, rich soil, or in wet ground such as is offered by a swamp or the edges of a pond. It is a good companion to the Cardinal-flower (*Lobelia cardinalis*), as they thrive under the same conditions and flower together.

HABIT OF GROWTH Upright, bushy plants to 3 feet or less.

FLOWERS Tubular, lavender-purple flowers about 1½ inches long, produced singly on long petioles in leaf axils from early summer to early fall. Each flower, like a Pansy, has a "funny face." Howard S. Bradley of Hamilton, New York, has introduced a white-flowered form.

FOLIAGE The leaves are opposite and broadly lance-shaped, and clasp the stems.

ASSETS The pretty purple-blue flowers appear at a welcome time. Monkey-flower, although considered a wet-soil plant, will do well in average garden soil so long as it doesn't become bone-dry.

FAULTS The flowers, although individually showy at close range, often blend with the foliage. The Monkey-flower is not for dry soil.

CULTURE Sun or light shade, rich or average soil that retains moisture, suits Monkey-flower. Plants are easy to transplant, and established clumps can be readily divided in early spring. It self-sows.

OCONEE-BELLS

TYPE Evergreen perennial
FAMILY *Diapensiaceae* (Dye-ap-pen-see-*ay*-see-ee)
GENUS *Shortia* (*Short*-ee-uh)
SPECIES *galacifolia* (gay-las-if-*foh*-lee-uh)
ORIGIN North and South Carolina mountains

This ground-hugging evergreen has a background somewhat reminiscent of another choice American native, the Franklin tree, *Gordonia alatamaha,* which was discovered in Georgia, introduced into colonial gardens and then never again found growing in the wild. Shortia's native range is restricted to a few areas in the Piedmont and in the Blue Ridge Mountains, so it is not surprising that it became "lost" to botanists after its initial discovery in 1781 by the French collector Michaux. About a hundred years later it was rediscovered, and is now fairly common in damp, acid-soil areas within its limited range. Fortunately most native plant nurseries propagate Shortia or Oconee-bells in great quantity.

USES As a ground cover in shade and companion plant to Rhododendron, Azalea and Mountain-laurel. Clumps can also be effective in rich-soil pockets in shaded, naturalistic rock gardens and at the base of low walls.

HABIT OF GROWTH Low-growing clumps 3 to 6 inches high, spreading by underground roots.

FLOWERS Bell-like white or pink-tinged flowers on short stems in mid-spring. Each stem produces but one flower, about an inch across, with five petals attractively fringed.

FOLIAGE Shining and waxy-textured evergreen leaves, 1 to 3 inches wide, oval in shape with notched and wavy margins. In late fall the leaves become reddish and bronze at their margins, and by winter are usually completely bronze.

ASSETS Lovely in flower in early spring and handsome all the year because of its evergreen leaves. Shortia is readily established in shady, acid-soil wild gardens.

FAULTS Shortia requires acid soil.

CULTURE Buy plants of Shortia in early spring or fall from native plant specialists. Plant them in shade in acid soil containing leafmold and/or peat moss. The soil should be well drained but retain moisture; otherwise the plants make only stunted growth. While the plants can stand some filtered sunlight, too much exposure is undesirable, one result being a yellowish color in the leaves instead of a rich green. After being established for a few years, the clumps can easily be separated after flowering or in early fall.

PARTRIDGE-BERRY

TYPE Evergreen sub-shrub
FAMILY *Rubiaceae* (Roo-bee-*ay*-see-ee)
GENUS *Mitchella* (Mit-*chell*-uh)
SPECIES *repens* (*ree*-penz)
ORIGIN Nova Scotia to western Ontario, Minnesota and Arkansas, south to northern Florida and Texas

Known to everyone is the Partridge-berry. It is a popular subject for terrariums and around Christmas time, for rounded glass bowls which are filled with the plants and tied with a red bow. The Partridge-berry seems uniquely well suited to such treatment: a few berried sprays stuffed into a small, closed pill bottle were sent to the *Flower Grower* office by Robert C. Baur around Christmas and eleven months later, still in their bottle, were in good condition. Other names for Partridge-berry are Twin-berry and Squaw-berry.

USES A dainty yet solid ground cover for woodlands, large patches being formed under good conditions. It is also suited to shaded, naturalistic rock gardens. As mentioned above, it is a long-lasting plant for terrariums and the like.

HABIT OF GROWTH Ground-hugging plants up to 3 inches or so high, with trailing, creeping stems a foot or more in length. It is evergreen and old stems are woody or semi-woody.

FLOWERS The fragrant white or pink-tinged flowers are produced in pairs from the leaf axils in late spring. Each flower, about ½ inch long, is tubed, and the petals appear fringed because of the beard inside. The red berries which follow are bright red, about ⅓ inch across, and are long-lasting. Occasionally a form with white berries is encountered. An interesting point is that but one berry is formed from the two flowers.

FOLIAGE Neat, rounded evergreen leaves about ¾ inch long with white veins.

ASSETS One of our most appealing ground covers, its diminutive appearance hiding its rugged constitution. The long-lasting berries show brightly against the evergreen leaves. Plants spread rapidly under good conditions.

FAULTS Not a good choice for poor, dry woodlands.

CULTURE Shade and acid soil, moderately moist and rich in humus, is the preference of Partridge-berry. In dry woods it will grow, but never as happily as in cool, rich woodland. Partridgeberry is not too difficult to dig from the wild if mats are carefully lifted, planted in well-prepared soil containing leafmold and/or peat moss and watered as necessary until the roots have become established. Cuttings of new growth can be taken in midsummer, inserted in flats of peat moss and sand. The berries can be planted in the fall and should germinate in the spring. Most wild-flower specialists offer plants which will easily settle down in the home garden if some attention is paid to the requirements already given.

PARTRIDGE-PEA

TYPE Herbaceous annual
FAMILY *Leguminosae* (Leg-yew-min-*noh*-see)
GENUS *Cassia* (*Kass*-ee-uh)
SPECIES *fasciculata* (fas-sik-yew-*lay*-tuh)
ORIGIN Wide distribution in North America

Sunny, sandy fields are brightened in midsummer by this Partridge-pea, also known as Wild Senna or Large-flowered Sensitive Pea. (In the same genus is *Cassia marilandica,* Wild Senna, which is a tall perennial up to 4 feet, bearing showy clusters of yellow flowers in summer. Its range is the Eastern United States and it grows under the same conditions as the Partridge-pea.)

USES The Partridge-pea can be grown in any dry, sunny or lightly shaded areas and is especially useful in sand-dune areas, on banks or slopes or in rock gardens.

HABIT OF GROWTH Low-growing, sometimes sprawling plants from 8 inches or so, but they may reach 1 or 2 feet.

FLOWERS From a distance the flowers gleam like drops of gold. Each flower, about ¾ to 1 inch across, has five golden petals touched with purple-red at the base and borne at the leaf axils. Long seed pods follow the flowers.

FOLIAGE The delicate, bright-green leaves are alternate and are composed of many leaflets which may fold together when handled.

ASSETS The bright flowers produced over a long period from early summer into fall; the tolerance of the plant for drought and sandy, poor soil.

FAULTS The seed is rarely offered commercially.

CULTURE The Partridge-pea and its related species, Wild Senna, should be grown from seed sown in early spring, or late fall for spring germination. Young plants can be carefully lifted for transplanting in spring when the soil is moist, although like most Legumes, they resent root disturbance.

PHLOX

TYPE Herbaceous perennial
FAMILY *Polemoniaceae* (Pol-em-oh-nee-*ay*-see-ee)
GENUS *Phlox* (Flox)
SPECIES *andicola* (an-*dik*-ol-uh), *bifida* (*bye*-fid-uh), *divaricata* (div-ar-ik-*kay*-tuh), *maculata* (mak-yew-*lay*-tuh), *ovata* (oh-*vay*-tuh), *pilosa* (pye-*loh*-suh), *rigida* (*rij*-id-uh), *stolonifera* (stoh-lon-*nif*-er-uh) *subulata* (sub-yew-*lay*-tuh)
ORIGIN See Species, below

It often comes as a surprise to realize that all the various forms of Phlox that predominate in cultivated gardens from spring to summer are North American in origin. The Phlox genus is a varied and versatile one which includes many alpine, low-growing species suitable for scree and rock-garden situations as well as species that inhabit rich woodlands. In between these two extremes, which may nonetheless overlap, are several species common to meadows and thickets. *Phlox paniculata* belongs to the latter category and, as almost every gardener knows, is the species whose numerous named varieties are the giants (in both quality and size) of perennial borders in summer. In fact, this species is a good example of how hybridization and skillful selection can improve on nature, for in the wild *P. paniculata* is not especially outstanding. Not so with many other Phlox species, however, which possess an informal and charming habit that does not cry out for improvement. A few of these species are included here.

USES The Phlox genus abounds with good plants for the home gardener. There are species suitable for dry, sunny slopes (*Phlox bifida* and *P. subulata*), for woodland and shaded sites with good soil (*P. divaricata*), for sunny rock gardens with rich soil pockets (*P. andicola* and *rigida*) and for a variety of half-shaded situations where a self-sufficient, ground-covering plant is needed (*P. ovata, divaricata, stolonifera, subulata*).

HABIT OF GROWTH Mat-forming and spreading plants only a few inches or so high as well as clumps whose flowering stems may reach a foot or more.

FLOWERS Five-lobed, narrow-tubed flowers, roughly star-shaped, borne in clusters, in spring or summer depending on the species. Many are quite fragrant. The color is predominantly pink or magenta.

FOLIAGE There is great variation. The alpine and ground-hugging species like *Phlox rigida* and *subulata* have prickly, needle-like leaves, small and stiff, which are evergreen but nearly invisible when the plants are flowering, while the taller species like *P. divaricata* and *maculata* have quite prominent lance-shaped leaves.

ASSETS The Phlox genus is famous for its profusion of flowers, usually lasting over a long period. Many species are adaptable to a wide area and varying conditions. Others are especially suited to difficult situations, including sunny, dry banks and rocky slopes. At least one species, Blue Phlox (*P. divaricata*) is as useful in cultivated and formal flower borders as it is for naturalizing in informal woodlands and under tall trees. Actually there are one or more Phlox types suitable for every section of this country.

FAULTS Overplanting and misuse of some species have kept us from appreciating the many other worth-while, shyer members of the genus.

CULTURE In general, these plants are not difficult to establish, the important trick being to understand conditions under which the various species grow in their native habitats. To a reasonable degree, these conditions should be duplicated in the home garden. Both the Blue Phlox and the Creeping Phlox (*P. divaricata* and *P. stolonifera*) need rich, humusy soil which never becomes bone-dry, as well as light shade in summer. The Downy Phlox (*P. pilosa*) is suitable for naturalizing in dryer, open woodland. Both the Sand Phlox and Moss-pink or Moss Phlox (*P. bifida* and *P. subulata*) are tolerant of sun and lean, sandy soil. Most Phlox, once established, self-sow, and most are readily increased by division of the clumps in early spring or after flowering.

SPECIES *Phlox andicola.* Native to North Dakota and Colorado. Mat-forming plant with needle-like foliage, white flowers about ¾ inch across, in late spring. Lavender-blue selections are available from Claude A. Barr. Plant in good soil in pockets of a sunny or lightly shaded rock garden.

P. bifida. Sand Phlox. Native from Indiana and Michigan to Iowa, Tennessee and Oklahoma. Semi-woody plant to about 8 inches, with downy, narrow, stiff leaves, 1 to 2 inches long. Pale lavender flowers, each petal deeply notched, in spring. For dry situations, rock gardens and walls, in sun. Propagate by seeds or cuttings.

P. divaricata. Wild Sweet William. Blue Phlox. (Illustrated) Native over a wide area, from Quebec to Michigan and Nebraska, south to Florida and Texas. Lavender-blue flowers in terminal clusters on 8- to 12-inch stems, each clump producing several flowering stems, making a very pretty effect in the early to mid-spring wild garden. This species likes rich, humusy soil and thrives in sun or light shade. It is wonderful with spring bulbs, Ferns, Primroses and other native plants. Use it in the foreground of the perennial garden, along woodland paths, in the rock garden, in retaining walls and in front of informal stone walls. It is easily increased by dividing the clumps after flowering. Many nurseries offer white forms as well as improved variations of the type.

P. maculata. Native from Connecticut to North Carolina and Iowa. Pink or purple flowers in a rounded cluster, each flower about ½ inch across, in summer, on 6- to 18-inch purple-spotted stems. Needs rich, moist soil in sun or part shade.

P. ovata. Wild Pink Phlox. Native from Alabama to Pennsylvania. Pretty rose-pink flowers, 1 inch across, in clusters, in late spring, on 8- to 12-inch stems above semi-evergreen mats which spread by stolons. For rock gardens and walls, as ground cover in woodland.

P. pilosa. Downy Phlox. Wild Sweet William. Native over a wide area from Ontario to Florida and Texas. Purplish-pink flowers, ¾ inch across, in clusters on 1-foot stems in late spring. Prefers moderately acid soil with some humus, and sun or light shade.

P. rigida. Sand-hill Phlox. Native from Montana to Washington. Alpine species worth trying where summers remain cool. Mat-forming, with sharp, needle-like foliage which is covered when the plants are in bloom. Flowers, about ½ inch across, in spring, their color varying from lavender-blue and purple to pink and white, depending on the acidity of the soil. Grow in rock gardens and retaining walls in sun.

P. stolonifera. Creeping Phlox. Native from Pennsylvania to Georgia. An attractive creeping species which produces 6-inch, erect stems bearing small clusters of mauve-pink flowers, 1 inch across, in spring. There are at least two named selections available: "Blue Ridge," with lavender-blue flowers, and "Pink Ridge," with pink flowers. Use as a ground cover in woodlands, along shaded paths, in the foregrounds of naturalized shrub groupings of Azalea, Rhododendron and the like and in rich soil pockets of shaded rock gardens. It needs humusy soil that retains moisture. Clumps are easy to divide.

P. subulata. Mountain-pink. Moss Phlox. Moss-pink. Native from New York to Michigan, south to North Carolina and Tennessee. This well-known, much-planted species in its rainbow of flamboyant colors has become the darling of hybridizers and may seem out of place among other natives. Yet even nature has had its fling with this species, as flowers in the wild state vary in color from magenta to rose, pink and white. It takes a good deal of color sense to use this so-useful Phlox in the home garden. It will readily carpet sunny, dry, steep slopes and banks where few other plants will take hold, and will soften the raw edges of misplaced stones and boulders. Sometimes, though, when color combinations have not been worked out well, the cure seems worse than the problem!

PIPSISSEWA

TYPE Evergreen sub-shrub
FAMILY *Pyrolaceae* (Pihr-ol-*ay*-see-ee)
GENUS *Chimaphila* (Kye-*maf*-fil-uh)
SPECIES *maculata* (mak-yew-*lay*-tuh), *umbellata* (um-bel-*lay*-tuh)
ORIGIN See Species, below

A choice little evergreen for dry woods is the Pipsissewa, also called Prince's-pine. Striped or Mottled Pipsissewa (*Chimaphila maculata*), which is illustrated, is especially handsome because its leaves have white veins. (Close relatives of Pipsissewa are Indian Pipe (*Monotropa uniflora*) and Pine Sap (*M. hypopitys*), which have bewitched generations of nature lovers. These curious, naked-appearing plants which spring up from layers of leafmold in summer, lack chlorophyll, the ingredient that gives the green coloring to other plants. While clumps can often be carefully lifted and transplanted to the wild garden, the trick is to get them to reappear the following year.)

USES Pipsissewa is for dry woodland and shaded rock gardens. It is too slow-growing to make a dense ground cover, but clumps make attractive accents under trees. The plants show up especially well against or among large stones or Pine needles. It is a popular subject for terrariums.

HABIT OF GROWTH Erect, low-growing plants about 6 inches tall, with semi-woody stems and spreading by underground shoots.

FLOWER Delicately sculptured flowers appear in a loose cluster at the top of the stems. Each flower, slightly scented and white or pink-tinged, droops slightly, has five petals and is about ½ inch across. Flowering time is late spring, although some plants produce flowers much later.

FOLIAGE Opposite, toothed leaves of a leathery texture, usually in whorls or clusters around the stem. They are evergreen.

ASSETS These are appealing plants, attractive all winter because of their evergreen foliage. They are tolerant of shade and dry soil and are decidedly worth planting in groups in dry woods to form a sparse ground cover. The flowers are charming. Most wild-flower nurseries carry plants.

FAULTS Difficult to transplant because of their straggly, far-reaching roots.

CULTURE Dry, acid soil in a woodland, preferably near and about Pine trees and other Conifers, suits the Pipsissewa. Under such conditions, the plants persist, slowly spreading by their underground stems and seeds. Plants can be moved from the wild, but as mentioned above, it is not as easy as the neat little clump seems to suggest. Nurseries usually propagate their plants from stem cuttings inserted in a mixture of sand and peat. The resulting plants have more compact root systems, making replanting a more successful operation.

SPECIES *Chimaphila maculata*. Mottled Pipsissewa. Spotted Wintergreen. Native from Maine and Ontario to Michigan and Illinois, south to Georgia and Tennessee. The more choice of the two species because of its handsome, variegated leaves which can be especially conspicuous during the winter. Each pointed leaf has white veins. Its flowers are slightly fragrant.

C. umbellata. Pipsissewa. Prince's-pine. Green Pipsissewa. Its range is extensive, from Nova Scotia to British Columbia, down to Georgia and then westward to the Rocky Mountains. Plants reach a height of 10 inches or so but the average is lower, about 4 inches. The toothed, evergreen leaves are bright green and from 1 to 2½ inches long, in clusters or whorls around the stem.

PITCHER-PLANT

TYPE Herbaceous perennial
FAMILY *Sarraceniaceae* (Sar-ras-seen-ee-*ay*-see-ee)
GENUS *Sarracenia* (Sar-ras-*seen*-ee-uh)
SPECIES *flava* (*flay*-vuh), *purpurea* (pur-*pew*-ree-uh)
ORIGIN See Species, below

The Pitcher-plant, with its rosettes of water-holding leaves, is one of the most intriguing of native plants. The purpose of these leafy receptacles is to catch and drown insects, the more palatable parts of the insect (to the Pitcher-plant, that is) being eventually digested by the leaves. The Pitcher-plant and its Far Western relative, California Pitcher-plant (*Darlingtonia californica*), are both members of the same family, but also of interest is yet another family (*Droseraceae*) of carnivorous plants to which the Sundews and the Venus Fly-trap belong. Rather than enticing insects to their leaf parts and then drowning them, as do the Pitcher-plants, the Sundews and Venus Fly-trap attract insects by means of small hairs on their leaves, each hair being dotted by a drop of sticky, sweet sap. The hairs actually fold over the insect which, by means of further plant secretions, is broken down so that the plant can digest it. Fantastic though this process may seem, it has been proved by the skeletal remains of the captured insects. All of these insectivorous plants naturally inhabit bogs and swamps and wet sandy soil that is acid. The Sundews (*Drosera*), which often carpet an entire sphagnum bog and are prolific with their seeds, can be scooped gently out of the moss and transplanted into a homemade bog garden that is acid. They are tiny, dainty plants, their rosettes of leaves, often reddish, sparkling as the sun catches the drops of fluid at the ends of their leaf hairs. The Venus Fly-trap (*Dionaea muscipula*) has been sadly exploited and, as its distribution is limited to sections of North and South Carolina, conservation-minded citizens of these states should help authorities who restrict the wholesale digging of these plants. Fortunately the Pitcher-plants—especially *Sarracenia purpurea,* which has wide distribution—are in good supply and should be grown in every home wild-flower garden if proper conditions can be provided.

USES The Pitcher-plant is ideal in natural or man-made bogs or in low, moist ground, especially around pools and streams.

HABIT OF GROWTH Its curious leaves and flowering stems grow directly from a creeping underground rhizome.

FLOWERS Nodding, cup-shaped flowers with five petals borne

singly atop a naked stem about one foot high, in spring and early summer.

FOLIAGE The radiating leaves—hooded, tubular and winged on one side—may be 4 to 10 inches long. Like a pitcher, they hold water in which insects are drowned, having been lured there by a secretion of the leaf and prevented from escaping by numerous hairs which line the inside of the leaves. (While the plants do utilize the dissolved insects, many botanists also believe that these remains attract and supply food to a fly's larvae, which assist in pollinating the flowers.) The leaves are handsomely colored, usually in tones of purple and green with deeper veinings.

ASSETS Interesting plants, handsome in foliage and flower.

FAULTS Not suitable for heavy shade or dry or average soil conditions.

CULTURE Pitcher-plants belong to peat and sphagnum bogs that are acid. In home wild-flower plantings they can be readily introduced into an artificial bog garden in which the growing medium is of acid peat moss and sand with sphagnum moss on top. Or plant them around informal pools where the soil remains moist, first mixing sand and peat in their planting holes, if these materials are otherwise lacking. Pitcher-plants will grow in light shade but prefer sun, and should receive at least a few hours of sunshine a day. The Pitcher-plant can be kept as a house plant, potted in a mixture of peat moss and sand with a layer of sphagnum moss on the top. The mixture should be kept moist but not soggy and not in strong sun all day. Winter temperatures ideally should be about 50 degrees. Once clumps are established, it is best not to interfere with them, although one way of propagating is by division of the rhizomes in spring.

SPECIES *Sarracenia flava.* Trumpet Pitcher-plant. Trumpet-leaf. Native Yellow Trumpet-leaf. Native from Virginia to Florida. A handsome species with slender chartreuse leaves 1 to 2 feet high and yellow flowers 2 to 4 inches across, appearing in late spring. Hardy north, or at least to New York City and Long Island.

S. purpurea. Pitcher-plant. (Illustrated) Native across Canada, south to Florida along the coast, inland to parts of Indiana, Illinois, Minnesota and the Rockies. Leaves, mottled and deeply veined, usually bronze or purple, about 8 inches long. Flowers, about 2½ inches across, may be purple or dark red, and appear in early summer.

PRICKLY-PEAR

TYPE Evergreen perennial
FAMILY *Cactaceae* (Kak-*tay*-see-ee)
GENUS *Opuntia* (Op-*punt*-ee-uh)
SPECIES *compressa* (kom-*press*-uh), *fragilis* (*fraj*-il-iss), *polyacantha* (pol-ee-ak-*kanth*-uh)
ORIGIN See Species, below

These Midwestern and Eastern representatives of the Far West's Cholla, or Prickly-pear, look like exotic aliens compared to the majority of natives associated with wild-flower gardening in temperate climates, but legitimate natives they are. They are useful and even attractive in certain situations.

USES In rock gardens, in rock-strewn situations or on slopes and banks, or in dry, sun-drenched seashore gardens.

HABIT OF GROWTH Sprawling, flat-stemmed clumps, each stem composed of Pear-shaped, spine-covered joints (called "pads") which leave no doubt that this plant is a member of the Cactus family. In addition to the spines, the joints are dotted with smaller bristles in bundles. Each plant has a system of wide-spreading horizontal roots as well as those that grow down vertically to reach moisture that can be stored in the fleshy joints.

FLOWERS Very showy and cup-shaped, usually yellow and about 2 inches or more across, with prominent stamens. They grow from the joints, in late spring, early summer and summer and are followed by fleshy, red-skinned fruits which are spineless, but are covered by the same vicious little bristles that are found on the joints. The fruits are edible, but watch out for those bristles.

FOLIAGE Leaves are hardly important, as they are small and are shed quickly.

ASSETS Their beautiful flowers and durability in drought and poor-soil situations.

FAULTS Care must be taken when working around these plants to avoid the bristles which come off easily and can become imbedded in the skin of arms, legs and hands.

CULTURE The Prickly-pear needs full sun and a neutral or just slightly acid soil that is well-drained at all times. The soil can be sandy or gravelly and lime should be added before planting if it is very acid. Propagation is easy by breaking off the joints or pads in spring and inserting them in sandy soil.

SPECIES *Opuntia compressa.* Prickly-pear. (This species is often listed as *Opuntia humifusa.*) Native from Ontario and Massachusetts to Alabama, west to Minnesota and Oklahoma. Stems about 5 inches high. Yellow, red-centered flowers, 2 to 3 inches across, in summer.

O. fragilis. Brittle Cactus. Native from Wisconsin to British Columbia and south to Texas and Arizona. Low-growing mounds to 3 inches or so, with large pale yellow flowers, 2 inches across, in early summer.

O. polyacantha. Prickly-pear Cactus. Native from North Dakota, Alberta, Washington, south to Texas and Arizona. Low-growing mounds with beautiful lemon-colored flowers, 4 inches across, in spring. Claude A. Barr has introduced some interesting color selections of this species, including orange, carmine and chartreuse.

RATTLESNAKE-PLANTAIN

TYPE Evergreen perennial
FAMILY *Orchidaceae* (Or-kid-*day*-see-ee)
GENUS *Goodyera* (Good-*yeer*-uh)
SPECIES *pubescens* (pew-*bess*-senz)
ORIGIN Widely distributed in Eastern North America, possibly farther

Charm rather than glamour characterizes this little Orchid which, superficially, bears little resemblance to more showy, hardy members of the same family. The genus has also been known as *Epipactis*. The "rattlesnake" in its common name refers to the pattern on the leaves. In fact, under the laws of the Doctrine of Signature, the leaves of Rattlesnake-plantain, when applied to a rattlesnake bite, were supposed to cure the victim, because the leaves resembled the snake! There are several other Goodyera species in North America, all quite similar. This is the species most often offered by specialists. It is usually called Downy Rattlesnake-plantain or Downy Rattlesnake Orchid, because of the soft hairs on the flowering stem.

USES As accents in a woodland, especially in company with Conifers, Rhododendrons and Ferns. A woodland rock garden where the stones have been artfully sunk so as to provide a dark background make an especially suitable situation for the Rattlesnake-plantain. Small plants may be lifted in the fall and used in terrariums where their patterned leaves provide a needed accent.

HABIT OF GROWTH Low-growing plants which spread by slender, creeping rhizomes, the rosettes of foliage almost hugging the ground.

FLOWERS Small, barely ¼ inch long, white or green-white, helmet-shaped flowers crowded on stalks from less than a foot to about 16 inches, in late spring to midsummer. Although the individual flowers are tiny, when viewed under a magnifying glass they do look like the Orchids they are. And the entire flowering stalk is effective, especially if a colony has been established.

FOLIAGE The rounded, pointed leaves, which form a low rosette, are decidedly handsome and, as they are evergreen, are effective all year round. Each leaf, about 3 to 4 inches long, has a prominent central white vein with the lesser veins etched in white, the effect resembling the markings of a rattlesnake's skin.

ASSETS Exquisite plants to appreciate for their perfection of detail. The foliage is always attractive and the flower spikes appear in late spring and summer when good flowering natives are rare. And lastly, the Rattlesnake-plantain is one of the easiest of our native Orchids to cultivate. Fortunately it is carried by most nurseries specializing in native plants. It is comparatively simple for the amateur to propagate.

CULTURE The Rattlesnake-plantain is found naturally in open woodlands of Pine and Oak. Give plants an acid soil full of leafmold and light or half, not dense, shade. The soil should be well drained, but moisture seems essential. In fact, they will thrive on the edge of a bog garden, where they can receive the overflow of water at their roots. The Rattlesnake-plantain can be propagated in spring or late summer by separating the rhizomes.

RUE-ANEMONE

TYPE Herbaceous perennial
FAMILY *Ranunculaceae* (Ran-nun-kew-*lay*-see-ee)
GENUS *Anemonella* (An-em-on-*nell*-uh)
SPECIES *thalictroides* (thal-lik-*troy*-deez)
ORIGIN Widespread in eastern half of North America, from New England, Ontario, to Florida, west to Minnesota and Kansas

This fragile-appearing wild flower shares characteristics with two other natives, the Anemone and Meadow-rue, as both its common name and botanical name indicate. Its flowers resemble those of the Wood Anemone but differ in that two or more are borne to a stem. Its foliage looks like that of the Meadow-rue.

USES The Rue-anemone is best appreciated in colonies in woodland gardens, along a meandering path, on shaded slopes or in rich soil pockets in a naturalistic rock garden.

HABIT OF GROWTH Upright clumps, from 6 to 9 inches tall, which grow from small tubers.

FLOWERS The flowers, about 1 inch across, have several white or pink-tinged sepals, and are carried in clusters above the foliage. There is a rare, fully double form. Blooming time is early to mid-spring.

FOLIAGE The compound leaves are gray-green and quite similar to those of Meadow-rue (Thalictrum).

ASSETS The Rue-anemone is one of our loveliest wild flowers and, despite its delicate appearance, it is durable and comparatively easy to establish.

FAULTS The plant goes dormant in early summer, so its place in the home wild-flower garden should be carefully noted.

CULTURE Rue-anemone needs rich, woodsy soil that retains moisture. The soil can be acid or close to neutral. A situation in light to dense shade is necessary. Plants can be collected, preferably in late spring after the flowers have faded, and can be increased by carefully separating the little clusters of tubers in late spring.

SAXIFRAGE

TYPE Herbaceous perennial
FAMILY *Saxifragaceae* (Sax-if-rag-*ay*-see-ee)
GENUS *Saxifraga* (Sax-*iff*-rag-guh)
SPECIES *michauxi* (mee-*show*-eye), *virginiensis* (vir-jin-ee-*en*-siss)
ORIGIN See Species, below

These two species are but a fraction of the vast number of Saxifrages scattered over North America, Europe and Asia, especially in alpine and arctic areas.

USES These two species, Michaux Saxifrage (*Saxifraga michauxi*) and Early Saxifrage (*S. virginiensis*), as well as most other plants of the genus, are subjects for rock gardens, walls and banks. Both of these species tolerate part shade.

HABIT OF GROWTH Panicles or clusters of small flowers on 3- to 18-inch stems above rosettes or mats of foliage close to the ground.

FLOWERS Star-shaped flowers, white or pink, about ¼ of an inch across, in loose panicles, in early spring (Early Saxifrage) or late spring (Michaux Saxifrage).

FOLIAGE Fleshy, sparsely toothed leaves in a rosette.

ASSETS Attractive plant habit and pretty, very early spring flowers in the case of Early Saxifrage. These Saxifrages are not difficult to establish.

CULTURE The Saxifrages need the presence of rocks, and they are especially effective when they appear to be barely lodged in cracks and crevices. The Early Saxifrage flowers at the same time as Wild-ginger, and the two make good companions in lightly shaded soil pockets of stone retaining walls or rock gardens. Actually the Saxifrage is tolerant of varying conditions and in nature is found in both dry and rather moist situations and in sun or part shade. Plants are readily divided in the spring and easily collected from the wild.

SPECIES *Saxifraga michauxi.* Michaux Saxifrage. (Some prefer the specific name of *leucanthemifolia* for this species.) Native in highlands of Virginia to Georgia. White or pinkish, starry flowers in late spring on 6- to 18-inch stems. Tolerates part shade.
S. virginiensis. Early Saxifrage. Rock Saxifrage. (Illustrated) Native from New Brunswick to Georgia and Tennessee and west to

Missouri and Minnesota. Slightly fragrant, star-shaped white flowers in early spring on a stem which varies from 4 to 12 inches in height, depending on moisture and exposure. A neat, undemanding native, most cheering in early spring.

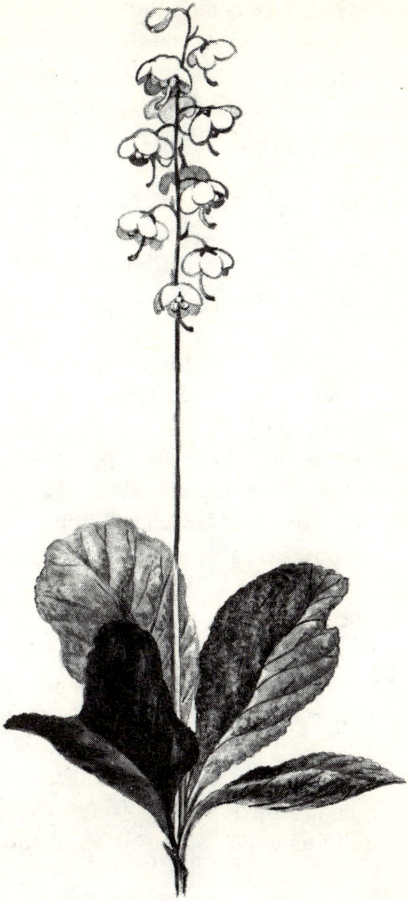

SHINLEAF

TYPE Evergreen perennial
FAMILY *Pyrolaceae* (Pirh-ol-*ay*-see-ee)
GENUS *Pyrola* (*Pirh*-ol-uh)
SPECIES *elliptica* (el-*lip*-tik-kuh), *rotundifolia* (roh-tun-dif-*foh*-lee-uh)
ORIGIN See Species, below

The Shinleaf or Pyrola has been shunted from one family to another by botanists, some placing it in its own family, *Pyrolaceae,* others in the closely related *Ericaceae.* At any rate, Pyrola is a near relative of Pipsissewa, which has similar growth habits, and the intriguing Indian Pipe and Pine Sap.

USES As occasional ground-cover accents in acid-soil woodlands, especially under Pines and Oaks and such shrubs as Rhododendrons and Azaleas, or in shaded, naturalistic rock gardens in prepared soil pockets.

HABIT OF GROWTH The plants spread by underground stems to form a sparse cluster or rosette of foliage.

FLOWERS The Shinleaf has nodding, white or pink-tinged flowers of a delicate, waxy texture about ⅔ of an inch across, carried in a raceme from 6 to 12 inches above the foliage, in late spring and early summer. The flowers are sweetly scented.

FOLIAGE Attractive, glossy evergreen leaves which are conspicuous until covered by fallen leaves in the fall. They generally are broadly oval, tapering toward the petiole, and are about 3 inches long, prominently etched by a network of veins.

ASSETS Their evergreen, shining foliage, attractive the year round, and the racemes of delicate flowers in early summer. They are interesting additions to any acid-soil woodland garden.

FAULTS Like their near relative, Pipsissewa, the Shinleafs appear deceptively easy to transplant from the wild, but their sparse, ranging roots are easily damaged or broken in the process. Also the plants need humusy, acid soil in shade.

CULTURE Some wild-flower nurseries offer plants of Shinleaf which should be ordered for spring or early fall planting. Plants should be set out in shade in an acid, humus-rich soil. Incorporate quantities of peat moss and/or leafmold in the planting holes. The species of Pyrola included here are fairly tolerant of dry soil, although all species do best in soil which retains moisture during drought. Newly set plants need careful watering at first. Propagation is by seed sown in peat moss and sand in fall for spring germination or by dividing the underground roots after flowering.

SPECIES *Pyrola elliptica.* Shinleaf. (Illustrated) Oval leaves tapering toward the base. Flowers are very fragrant, bloom in early summer. This species is native from Nova Scotia to British Columbia, south to Maryland, Illinois, Iowa and the Rocky Mountains to New Mexico.

P. rotundifolia. Round-leaved Shinleaf or Wintergreen. This species has been known as *P. americana* but is now considered the same as the European species. Its elongated, oval leaves are very handsome, as are its nodding, cup-shaped and fragrant flowers which appear in late spring. Native from Nova Scotia to South Dakota, south to Georgia and Ohio.

SHOOTING-STAR

TYPE Herbaceous perennial
FAMILY *Primulaceae* (Prim-yew-*lay*-see-ee)
GENUS *Dodecatheon* (Doh-dek-*kayth*-ee-on)
SPECIES *cusicki* (kew-*sick*-eye), *meadia* (*mee*-dee-ah), *pulchellum* (pull-*kell*-um)
ORIGIN See Species, below

The Shooting-star, once very prevalent in the prairie states, belongs to the Primrose family. In fact its flowers show a rather close resemblance to another member of the same family, the Cyclamen.

USES In woodlands, in rock gardens or groups in any lightly shaded site.

HABIT OF GROWTH Rosettes of foliage from a small rhizome with thick roots, the foliage usually low to the ground. The naked flowering stem may reach 6 inches or a foot or more in height.

FLOWERS The flowers have five petals that are swept backward and expose the anthers, which resemble a beak. The flowers, usually several in number, are clustered at the top of the stem, and droop. Blooming time is spring and late spring.

FOLIAGE Rosettes of lance-shaped, smooth leaves about 6 inches long, which disappear in summer.

ASSETS Exquisite flowers and a graceful plant habit make these choice natives for a spring garden.

FAULTS Mark the location of the plants well, as their foliage will have disappeared by late summer and fall.

CULTURE All the Shooting-stars prefer acid, humus-rich soil that retains some moisture in periods of drought. Either full sun or light shade is satisfactory. Plants should be set out in early spring. Clumps can be divided immediately after flowering.

SPECIES *Dodecatheon cusicki*. Rose Shooting-star. British Columbia to Alberta, south to Washington and Montana. Deep rose or lavender flowers in late spring. Height about 12 inches.

D. meadia. Shooting-star. American Cowslip. (Illustrated) Pennsylvania to Manitoba and Texas. Rose and white flowers with purple anthers, but many variations exist. Height about 1 foot, sometimes taller. Late spring.

D. pulchellum. Prairie Shooting-star. North Dakota and adjoining areas. Deep rose or lavender-rose and yellow flowers about 6 inches tall, in spring. Two selections introduced by Claude A. Barr include White Comet and Red Comet, with larger, deep rose flowers.

SHOWY ORCHIS

TYPE Herbaceous perennial
FAMILY *Orchidaceae* (Or-kid-*day*-see-ee)
GENUS *Orchis* (*Ork*-iss)
SPECIES *spectabilis* (spek-*tab*-il-iss)
ORIGIN New Brunswick and Ontario to North and South Dakota, south to Nebraska, Missouri, Kentucky and Georgia

This wide-ranging Orchid, as fascinating as it is beautiful, is one of two species native to North America, the other being the rare *Orchis rotundifolia.*

USES For a woodland garden in rich soil.

HABIT OF GROWTH Two leaves grow upright from a small tuber with fleshy roots, the flowering stem between the leaves reaching a height of 4 to 12 inches.

FLOWERS Each flower, part of a raceme, is about an inch long, has bright purple petals and sepals and a ½-inch lip that is white. Anywhere from two to eight or more flowers may be carried on the rather stout, fleshy stalk. The flowers possess some fragrance. Flowering time is spring.

FOLIAGE A pair of dark green, shiny leaves, about 8 inches long. Leafy bracts occur on the stem between the flowers.

ASSETS The beauty of the flowers and entire plant habit. Fortunately this is one of our more tamable Orchids, *if* a reasonable approximation of its native growing conditions can be provided.

FAULTS Not for the slapdash gardener.

CULTURE The Showy Orchis naturally grows in moist, humus-rich soil which may be neutral or moderately acid, usually in hardwood and Conifer forests. While it doesn't demand boggy soil conditions, it should be grown in soil that retains moisture and doesn't dry out in drought. It can be settled successfully in shaded sections of a rock garden, providing the soil pockets are rich, cool and deep. Occasionally newly set-out plants remain dormant for at least a season, so their locations should be well marked. Propagation—for those fortunate enough to arrive at this lofty stage—is by cutting apart the tuberous root, allowing one or more buds to each division.

SKULLCAP

TYPE Herbaceous perennial
FAMILY *Labiatae* (Lay-bee-*ay*-tee)
GENUS *Scutellaria* (Skew-tel-*lay*-ree-uh)
SPECIES *incana* (in-*kay*-nuh), *integrifolia* (in-teg-rif-*foh*-lee-uh), *ovata* (oh-*vay*-tuh), *serrata* (ser-*ay*-tuh)
ORIGIN See Species, below

Helmet-flower is another common name for this genus and both it and the name Skullcap are for the slight hump on the calyx of the flower. The seed capsule that follows the flower is also shaped like a hat or helmet. Skullcap belongs to the Mint family.

USES Group at least three plants of a species together in open woodlands, along partly shaded paths or in informal borders of other native plants, such as Cardinal-flower, or use in rich pockets of a lightly shaded rock garden or against a stone wall.

HABIT OF GROWTH Bushy, upright plants having stiff, square stems from 1 to 3 feet tall.

FLOWERS The blue-purple flowers, about an inch long, are composed of two lips, the upper lip being shaped like a hood or helmet. The flowers, depending on the species, are produced in short terminal racemes or in pairs from the leaf axils. Blooming time is late spring or midsummer, depending on the species.

FOLIAGE Rather coarse, opposite leaves, heart-shaped and toothed, which may be 4 inches long.

ASSETS The rich blue-purple color of Skullcap's flowers, which is especially appreciated in the summer-blooming species.

FAULTS The plants, though not invasive, are rather coarse in appearance.

CULTURE Generally the Skullcaps prefer moist, rich soil in light shade. However, *Scutellaria ovata* and *S. incana* are usually found in open woods in rather dry soil. All Skullcaps can be increased by division in early spring or fall and, once established, will self-sow.

SPECIES *Scutellaria incana.* Downy Skullcap. Native from New York to Iowa and Kansas, south to Florida, Alabama and Arkansas. Blue flowers 1 inch long in a terminal raceme in summer. The undersurfaces of the leaves are whitened by a fine covering of hairs. Useful in rock gardens or flower borders in full sun or light shade. Average soil.

S. integrifolia. Larger Skullcap. Hyssop Skullcap. Massachusetts to Missouri, south to Florida and Texas. Bright purple-blue flowers, about 1 inch long, in terminal racemes, in early summer. Light shade and average, woodsy soil.

S. ovata. Native from Maryland to Minnesota, south to South Carolina and Alabama and west to Texas and Mexico. Blue flowers, accented by a white lower lip, bloom in summer. A pink selection has been introduced by Woodland Acres Nursery. This species tolerates sun or light shade and average soil but will perform well in rich, moist soil.

S. serrata. Showy Skullcap. Native from Ontario to New York, south to South Carolina and Tennessee, west to Missouri. Blue-purple flowers in pairs from the leaf axils at the ends of the stems in late spring. Each flower is about an inch long. The oval leaves are conspicuously toothed. Light shade and rich, acid soil.

SKUNK-CABBAGE

TYPE Herbaceous perennial
FAMILY *Araceae* (Ar-*ray*-see-ee)
GENUS *Symplocarpus* (Sim-ploh-*karp*-us)
SPECIES *foetidus* (*fet*-id-us)
ORIGIN Quebec to Manitoba, south to Virginia, Ohio, Illinois and Iowa

Without a doubt, the Skunk-cabbage offers the first, definite sign that spring will come. It is the first plant to push through the ground, even through snow and ice, and its exotic, hood-encased flowers are certainly the earliest of the natives considered in this book. A near relative is the much-later-appearing Jack-in-the-pulpit.

USES Well worth planting in rich, moist or wet woods and beside large pools or ponds (either man-made or natural) for its early appearance in spring and later for the striking accent value of its large foliage.

HABIT OF GROWTH The Skunk-cabbage has fleshy roots which go deep into the soil. The flowering parts precede the leaves.

FLOWERS The flowers of the Skunk-cabbage are small and inconspicuous and dot the nearly round or club-shaped stem which is almost hidden by the 3- to 6-inch, hood-like bract, technically a spathe. It is the spathe, made conspicuous by its attractive mottling of purple, green and brown, that punctuates otherwise dormant swamps, marshes and wet lands in late winter and early spring. Later a cluster of red berries appears, the flowers having been pollinated by carrion flies which have been attracted to the plant by its odor and coloring.

FOLIAGE The leaves follow the flowers, first appearing as a tightly furled cone which may push through the soil as early as March. As the leaves develop, they cause the plants to look like a Cabbage, although by summer this resemblance passes, as each leaf, on a long stem, may be 1 to 3 feet long and a foot wide.

ASSETS Its early flowering, its handsome foliage and tolerance for shaded, wet, poorly drained soil as well as rich soil of average moisture content.

FAULTS The plants do have an unpleasant odor, most noticeable when the plants have been crushed. The plant is interesting enough so that most gardeners can ignore the odor. The plants are too large for restricted situations.

CULTURE The Skunk-cabbage prefers rich, wet soil that is acid but will do quite well in more average woodland or shady sites where the soil never becomes bone-dry. Plants can be dug from the wild but with difficulty. Choose a small plant and dig as deep as possible, otherwise the roots will be chopped off. Most wild-flower nurseries offer plants.

SOLOMON'S-SEAL

TYPE Herbaceous perennial
FAMILY *Liliaceae* (Lil-ee-*ay*-see-ee)
GENUS *Polygonatum* (Pol-ig-on-*nay*-tum)
SPECIES *biflorum* (bye-*floh*-rum), *commutatum* (kom-mew-*tay*-tum)
ORIGIN Eastern half of North America

Solomon's-seal is one of our most handsome natives and scores high for spring, summer and fall interest. The common name refers to pronounced scars left by previous shoots on the rhizomatous root. (See False Solomon's-seal for ways these two plants differ.)

USES A valuable addition to the home garden in a variety of situations. It can be combined with Hosta in a semi-shady border along a wall or walk and colonies are attractive in portions of a perennial border when shade is a problem. The Solomon's-seal is appropriate to any wild garden, is effective on slopes where the arching stems show to special advantage, and makes an ideal companion to broad-leaved evergreens like Mountain-laurel and Rhododendron. A few stalks can be cut and make striking, long-lasting foliage arrangements indoors.

HABIT OF GROWTH Solomon's-seal has a knobbly rhizome from which the arching stems appear in early spring, eventually to 1 foot to even 4 or 5 feet high, depending on the species and locale.

FLOWERS Yellow-green bells which hang from the leaf axils all up the stem, in mid- to late spring. Each bell is from ½ to ¾ of an inch long and usually appears in pairs or clusters. The flowers are followed by berries, green in summer but turning dark blue in fall, possessing the bloom and beauty of ripe Blueberries.

FOLIAGE The leaves, oval or oblong and decidedly pointed, clasp the stem in alternate tiers right to the tip. Each leaf, from 2 to even 6 inches long, may be so prominently veined as to make the leaf appear puckered. Color varies from light green to bluish green. A few nurseries offer variegated leaf forms, which may be derived from Asiatic or European species. Whatever their origin, they are attractive and worth mixing with the green-leaved kinds.

ASSETS Perhaps not the showiest of native plants but handsome in all its parts, the Solomon's-seal possesses a sculptured quality. The fall effect is unique when the dangling berries have turned blue and contrast with the leafy stems. This is an adaptable native, not difficult to collect and offered by all wild-flower nurseries.

FAULTS The greenish bells of its flowers are inconspicuous compared to many spring-blooming natives.

CULTURE Both species may be found in a variety of soils, ranging from very acid to near neutral, and moist to quite dry. They seem equally tolerant of sun or deep shade or half shade. Generally, though, they do best in the home garden in light or part shade, in moist, acid soil full of leafmold. Plants encountered in the wild should be lifted in early spring, just as the new shoots are breaking through the soil, or in fall. The berries can be planted in fall (remove the pulp) and should germinate in spring.

SPECIES *Polygonatum biflorum.* True Solomon's-seal. Small Solomon's-seal. (Illustrated) Native from New Brunswick to Michigan and Nebraska, south to northern Florida and west to Texas. Stems reach about 3 feet, leaves to about 4 inches or less. Flowers are usually carried in pairs and appear in late spring. A variant or very similar species is *P. pubescens,* which has a downy coating on the undersides of the leaves.

P. commutatum. Giant Solomon's-seal. Great Solomon's-seal. (This species is sometimes listed as *P. canaliculatum* or *giganteum.*) Native over a large part of North America, from New England west to Michigan and North Dakota, south to Georgia and Texas. This species is tall (up to 5 feet in very rich, moist soil) and stately. It flowers in late spring or early summer.

SPRING BEAUTY

TYPE Herbaceous perennial
FAMILY *Portulacaceae* (Port-yew-lak-*kay*-see-ee)
GENUS *Claytonia* (Klay-*toh*-nee-uh)
SPECIES *caroliniana* (ka-rol-in-ee-*ay*-nuh), *virginica* (vir-*jin*-ik-uh)
ORIGIN See Species, below

One of the delights of wild-flower explorations is finding a wide-spreading carpet of Spring Beauty in an open woodland in early spring. Equally charming representatives of this genus occur in the West and Pacific Coast regions too.

USES As a carpeting plant in woods interplanted with Daffodils, Violets, Virginia Bluebells. Smaller groupings can be tucked into soil pockets in a shaded rock garden or at the base of trees and shrubs.

HABIT OF GROWTH Wiry, succulent stems, from 5 to 10 inches high, from a tuberous root.

FLOWERS They are about ½ inch across, the petals white or rose-tinted, with deep rose veins, carried in racemes. Blooming time lasts for several weeks, usually starting in April and continuing into mid-May.

FOLIAGE Sparse, bladelike leaves, rather fleshy and twisted.

ASSETS A dainty, beguiling native and one of the easiest to grow. Once established, it will take care of itself.

FAULTS The absent-minded gardener may destroy plantings in summer when aboveground growth disappears.

CULTURE Plant the tuberous roots about 2 inches deep in spring or fall, or right after flowering. Plants are easily dug from the wild. Established colonies usually self-sow. The soil for Spring Beauty should retain moisture and be moderately rich, although the plant performs fairly well in average soil. Partially shaded or open woods are best. Robert C. Baur has reported that Spring Beauty is one of the easiest wild flowers to force into early flower indoors. He lifts clumps in the fall or early spring, places them in pots and keeps the pots in a cool, well-lighted window. The forced plants can later be returned to their garden setting.

SPECIES *Claytonia caroliniana.* Carolina Spring Beauty. Found over a wide area, from Canada south to Missouri and

Louisiana. Similar to the Common Spring Beauty but with broader leaves.

C. virginica. (Illustrated) Common Spring Beauty. Found from Ohio south to Georgia and Texas. (Western species include *C. rosea* of the Rockies and *C. lanceolata,* found in parts of California, Utah and Wyoming.)

STARFLOWER

TYPE Herbaceous perennial
FAMILY *Primulaceae* (Prim-yew-*lay*-see-ee)
GENUS *Trientalis* (Trye-en-*tay*-liss)
SPECIES *borealis* (bor-ee-*ay*-liss)
ORIGIN Canada, south to Virginia, Illinois and Minnesota

Starflower, often called American Starflower or May-star, is a neat little native, especially suited to carpeting the ground around Rhododendrons. Its generic name, *Trientalis,* means ⅓ of a foot, and refers to the height of the plant.

USES As a ground cover or colonizer for rich woods or shaded naturalistic rock gardens or outcroppings. A group can supply unexpected charm along a woodland path or when planted beneath Rhododendrons and Azaleas.

HABIT OF GROWTH The Starflower's effect is delicate, almost fragile, its threadlike stem, rising from a creeping root, from 4 to 9 inches tall, carrying a single whorl of leaves about midway.

FLOWERS White flowers, ½ inch across, have very pointed petals, and appear at the top of the slender stem in mid- to late spring. The number of flowers borne by each stem varies from one to four, although the usual habit is to produce one or two.

FOLIAGE The leaves, from five to ten, are arranged in a whorl near the top of the stem, making a distinctive star shape when viewed from above. Each leaf is slender and pointed, from 1½ to 5 inches long.

ASSETS The Starflower offers unique charm, rather than great show, and dependability, despite its fairly dainty habit. Once introduced, the plants will spread to form an airy rather than dense carpet. The flowers, which may last for several weeks, are very pretty to look down on and seem suspended over the foliage whorls. Plants can be moved from the wild in spring or fall.

FAULTS Not a subject for poor, dry woods.

CULTURE Grow the Starflower in shade, in acid, humus-rich soil which retains moisture. (The plants will also thrive in shaded bogs.) Most wild-flower nurseries offer plants for spring or fall planting. Established colonies can be spread by dividing the underground, creeping stems. Or sow seed in late fall for spring germination or in spring in a sand-peat moss or leafmold mixture.

WILD FLOWERS CAN BE ENJOYED IN MANY WAYS, IN MANY PLACES

The grassy path, above, winds through a mixed planting of evergreen and deciduous trees and shrubs, with ample space on each side of the path for Ferns, Solomon's-seal and other plants. New homeowners can plan such areas according to space. Bloodroot (left), photographed by B. Cory Kilvert, Jr., in his garden, has been planted among Ivy.

MAKING A BOG GARDEN

The addition of an artificial bog garden makes it possible to enjoy unusual plants that need more than ordinary soil moisture. The author's bog garden is situated in a very dry, open woodland. A fallen log was utilized as one side of the bog which was made in a rectangular shape, about 7 feet by 4 feet and excavated to 2 feet. It was then lined with two layers of heavy polyethylene sheets and filled with peat moss and acid, sandy soil. A drainpipe to carry off surface moisture proved unnecessary, as mice made several holes in the plastic. Variations on man-made bogs include tubs and concrete-lined excavations, in which case drainage pipes are necessary.

The author's finished bog two days after construction, showing the old log which lined one side. First plants set out were Skunk-cabbage, Bottle Gentian and Pitcher-plant.

The Yellow Fringed Orchid blooms in summer and is well suited to a bog garden.

Another native Orchid, Rattlesnake-plantain, was successfully established at the edge of the same bog garden. Other plants which have done well include Purple Fringed Orchid, Cardinal-flower, Great Blue Lobelia, Goldthread and Sundew.

VEIN-LEAF HAWKWEED brightens acid, rather dry woodland with its yellow flowers in summer. Its leaves are in rosettes close to the ground.

SWAMP BUTTERCUP

TYPE Herbaceous perennial
FAMILY *Ranunculaceae* (Ran-nun-kew-*lay*-see-ee)
GENUS *Ranunculus* (Ran-*nun*-kew-lus)
SPECIES *septentrionalis* (sep-tent-ree-on-*nay*-liss)
ORIGIN New Brunswick to Manitoba, Georgia and Kansas

There are many Buttercups, both native as well as introduced species. All spread at a mad rate and unless one has the acreage of a farm, most Buttercups, cheering though they can be, had best not be intentionally introduced. Probably a few plants will turn up anyway. The Swamp Buttercup, which is sold by most wild-plant specialists, is an especially handsome species, valuable for certain situations.

USES A good ground cover for part shade and moist soil.

HABIT OF GROWTH Fairly clumpy and upright in the spring, with fleshy, smooth and hollow stems, which after flowering become ground-hugging. Each plant produces many long, procumbent stems all summer which root at the nodes.

FLOWERS Bright yellow flowers, 1 to 1¼ inches across, with five petals and many short stamens, in clusters at the ends of the stems. Flowering period is a long one, extending from early spring to early summer.

FOLIAGE The shiny, deep green foliage is impressive—coarse yet handsome, each leaf deeply lobed and divided into three segments, each with its own stem.

ASSETS A quick-spreading plant with good foliage and bright flowers which can be useful where space is ample.

FAULTS As with all the Buttercups, the Swamp Buttercup can be invasive and shouldn't be introduced into small-sized, mixed-wild-flower gardens with rich soil, as one spring-planted clump will have grown in leaps and bounds by fall. The Swamp Buttercup will not thrive in poor, dry soil.

CULTURE Rich, moist soil that is even wet suits this Buttercup, which is often found in bogs and swamps. It will grow in sun or part shade and is easily increased by transplanting the young plants which appear wherever the stems come in contact with the soil long enough to send down roots.

TRILLIUM

TYPE Herbaceous perennial
FAMILY *Liliaceae* (Lil-ee-*ay*-see-ee)
GENUS *Trillium* (*Trill*-ee-um)
SPECIES *cernuum* (*ser*-new-um), *erectum* (ee-*rek*-tum), *grandiflorum* (gran-dif-*floh*-rum), *luteum* (*lew*-tee-um), *nivale* (niv-*vay*-lee), *sessile* (*sess*-i-lee), *stylosum* (stye-*loh*-sum), *undulatum* (un-dew-*lay*-tum)
ORIGIN See Species, below

The Trilliums keep pace with spring's progress. Some, like Red Trillium (*Trillium erectum*), illustrated, right, which is most common in the East, and Wake-robin (*T. ovatum*), which inhabits Pacific coastal ranges, flower in early spring, while the Rose Trillium (*T. stylosum*) and the Painted Trillium (*T. undulatum*) bloom in spring's waning days as summer is rushing in. While many Trilliums are beautiful, other species are more curious than beautiful; yet the entire genus, which spreads all over North America, is fascinating. Trilliums are readily recognized because their leaves and flower petals are in threes. The common name of Wake-robin, though often used for the whole genus, is more often applied to the earliest-flowering species in any region for obvious reasons.

USES For woodland gardens, for rockeries, for partly shaded sections of perennial borders and for foreground planting in front of trees and shrubs.

HABIT OF GROWTH Upright plants, 6 to 24 inches tall, from thickened roots or tubers.

FLOWERS The distinctive flower is composed of three white or colored petals and three green sepals.

FOLIAGE The leaves, also in threes, are in whorls, each leaf deeply veined, broad and pointed, some with stems, others clasping the main stem. The Toad Trillium (*T. sessile*) has mottled foliage.

ASSETS The genus is most intriguing. There is hardly a section of the country where some Trilliums don't occur naturally. Collecting Trilliums from all regions can be a rewarding pastime, as many variants exist and all, with reasonable care, seem to thrive well beyond their original areas. The flowers are generally long lasting. Some kinds of Trillium are in bloom throughout the spring season.

FAULTS The usual problem with rhizome-rooted natives exists with Trilliums: rodents may devour the roots.

CULTURE Trilliums, while generally found in woods, do not necessarily require deep shade if their major soil requirements are met. Most need rich soil, acid or neutral depending on the species, that does not become bone-dry. The Painted Trillium (*T. undulatum*) is especially fussy, requiring very moist, acid soil, rich in humus and cool in summer. The White Trillium (*T. grandiflorum*), a truly great native plant, is one of the least demanding and will thrive for years in a partly shaded portion of a perennial border. Seeds of some Trilliums are always available from specialists and should be sown in the fall. Old clumps can be divided after flowering. When planting, the rhizomes should be set so they are covered by about 2 inches of soil. For the best effects, use at least six of a kind to a group.

SPECIES *Trillium cernuum.* Nodding Trillium. Native from Newfoundland to Wisconsin, south to Georgia and Alabama. The large leaves nearly hide the drooping white or pink-tinged flowers which appear in mid-spring. Plant height varies from 8 to 18 inches. The bright red seed is of interest in late summer. This Trillium needs rich, damp soil. Plant near a shaded pool so that the roots receive overflow from the pond or in a low area where the soil never becomes dry.

T. erectum. Red Trillium. Purple Trillium. Wake-robin. Birthroot. (Illustrated, right) Found from Quebec to Michigan, south to Georgia and Tennessee. Maroon flowers on short stems, usually upright but sometimes bent, in early spring. Height, from 10 to 15 inches. This is one species that offers much variation in petal color, as some forms may be more brown than maroon or purple, or yellow and near white. One nursery offers variety album, which has white or pinkish flowers. Grow in humus-rich soil.

T. grandiflorum. White Trillium. Snow Trillium. (This species is illustrated, left.) Wake-robin. Quebec and Maine to Minnesota, south to Georgia and Arkansas. Just about the best of the Trilliums and one of our most popular native plants. Large pure-white flowers, from mid- to late spring, which may be 3 inches wide, fade to pink. Height of the plant is from 8 to 18 inches. Beautiful with Azaleas, Rhododendrons and Ferns. It performs best in rich acid or neutral soil that doesn't dry out.

T. luteum. Yellow Trillium. Lemon-yellow form of *T. sessile* (see below).

T. nivale. Dwarf White Trillium. Pennsylvania to Kentucky and Nebraska. Small white flowers on 6-inch stems, in early spring, both the flowers and foliage having stems. Ideal for rich soil pockets in the partly shaded rock garden.

T. sessile. Toad Trillium. Toad-shade. New York to Missouri, south to Georgia and Alabama. Interesting plant with dark red flowers above mottled foliage, the leaves being stemless. Flowers in early spring. Variety luteum has yellow flowers. Height, 1 foot or less. Another good Trillium for rock gardens in light shade. Soil should be rich in humus but well drained, and slightly acid or neutral.

T. stylosum. Rose Trillium. North Carolina to Georgia and Alabama. Lovely species with drooping flowers on short stems in late spring. Plant height is from 10 to 18 inches. Give it light shade and rich woodland soil.

T. undulatum. Painted Trillium. Quebec to Manitoba, south to Georgia, Tennessee and west to Wisconsin. Another beautiful species this, with wavy-edged petals which are white, their centers stained or painted with rose. The flowers, 1½ inches across, are slightly drooping or held erect on short stems. Plant height is from 8 to 15 inches. Late spring is its flowering season and it needs shade as well as a rich, acid soil that remains reasonably moist.

TROUT-LILY

TYPE Herbaceous perennial
FAMILY *Liliaceae* (Lil-ee-*ay*-see-ee)
GENUS *Erythronium* (Ehr-ith-*roh*-nee-um)
SPECIES *albidum* (*al*-bid-um), *americanum* (am-eh-rik-*kay*-num), *californicum* (kal-if-*forn*-ik-um), *grandiflorum* (gran-dif-*floh*-rum), *hendersoni* (hen-der-*son*-eye), *mesachoreum* (mess-ah-*kor*-ee-um), *oregonum* (or-eg-*go*-num)
ORIGIN See Species, below

Other names for Trout-lily include Adder's-tongue, Dogtooth-violet and Leopard-lily, and in the West, Fawn-lily and Glacier-lily. While the Trout-lilies are known for their pretty mottled or spotted leaves, not all species possess marked leaves.

USES Plants for a spring woodland garden or a shaded rockery. They are especially effective in large groups at the base of trees or among Ferns and other spring-flowering members of the Lily family.

HABIT OF GROWTH The Trout-lily's root system consists of a corm or bulb, several inches deep, which forms rhizomes or offshoots. The aboveground plant has a pair of basal leaves and a flowering stem from 6 inches up to a foot or so, depending on the species. The plants lose their foliage in summer.

FLOWERS Nodding, Lilylike flowers, bell-shaped and with flaring petals, usually about 2 inches long, in early to mid-spring.

FOLIAGE Handsome, rather fleshy, shiny foliage, often mottled with brown. It disappears in summer.

ASSETS Attractive flowers and foliage which add enchantment to any woodland, large or small.

FAULTS Trout-lilies tend to be sparing with their flowers, and this seems to be true even when large colonies of them are found in rich woods. They are difficult to dig, as their roots are often quite deep. Plants, either collected or purchased from specialists, are often slow to flower, although this varies with the species. The Common Trout-lily needs rich soil that retains moisture and some shade.

CULTURE Plant Trout-lily bulbs in the fall, about 3 or 4 inches deep and 4 inches apart, and make groups of at least six of a kind. While they like a rich soil, acid or neutral, that retains moisture, they do need good drainage. Mix leafmold and/or peat moss in the soil before planting, and if the soil is very sandy, add some clay to help retain moisture. The Common Trout-lily and a few other species form offsets from their bulbs which can be removed to form new plants. Most of the Far Western Trout-lilies will thrive with reasonable success elsewhere. Several of these are listed below.

SPECIES *Erythronium albidum.* White Fawn-lily. Ontario to Minnesota, south to Kentucky and Oklahoma. White flowers in early spring on 8-inch stems between a pair of mottled leaves. This species produces offsets which come into flower faster than those of *E. americanum.*

E. americanum. Common Trout-lily. Eastern Trout-lily. (Illustrated) Nova Scotia to Minnesota, south to Florida and Oklahoma. Yellow flowers, sometimes touched on the outside with purple or brown, in spring. Mottled foliage. The offsets from this species' bulbs are especially poky about reaching flowering size. However, the spotted leaves are pretty even without the flowers.

E. californicum. Fawn-lily. Native to California. Creamy-white flowers above very mottled foliage, the stems sometimes to 12 inches. Early spring.

E. grandiflorum. Glacier-lily. British Columbia, to Oregon and Utah. Bright yellow flowers on 12-foot or taller stems above unmottled leaves in spring.

E. hendersoni. Henderson's Trout-lily. Southern Oregon. Pale lavender flowers with dark-purple centers on 10-inch stems above mottled leaves in early spring.

E. mesachoreum. Midland Fawn-lily. Native to the prairie states and perhaps a variant of *E. albidum.* White flowers with lavender tints above plain foliage. The bulbs do not form rhizomes or offsets. This species will tolerate more sun than most other Trout-lilies but still needs rich soil. It blooms in early spring.

E. oregonum. Fawn-lily. Chamise-lily. British Columbia to northern California. Usually listed as *E. giganteum.* Straw-colored or creamy petals on 10-inch stems above mottled foliage in spring. A beautiful plant in its native habitat which does quite well elsewhere in rich soil.

TURTLE-HEAD

TYPE Herbaceous perennial
FAMILY *Scrophulariaceae* (Skrof-yew-lay-ree-*ay*-see-ee)
GENUS *Chelone* (Kee-*loh*-nee)
SPECIES *glabra* (*glay*-bruh), *lyoni* (*lye*-on-eye)
ORIGIN See Species, below

The flowers of Turtle-head actually look like a grinning turtle, or at least a turtle with his mouth slightly agape. The Turtle-heads belong to the Snapdragon family.

USES Plant in damp soil, especially where late summer flowers will be appreciated. The Turtle-heads make good companions for the Great Blue Lobelia as well as the Cardinal-flower, Bottle Gentians and Black Snakeroot. They are effective around natural or man-made pools and thrive in bogs and swamps. A group of Turtle-head plants is not out of place in a cultivated flower border, especially where there is light shade and moist soil.

HABIT OF GROWTH Neat, bushy clumps which produce one or several tall flowering stems which vary from 1 to 4 feet.

FLOWERS Tuberous, two-lipped flowers, about 1 inch long, in close spikes at the ends of the stems, in summer and early fall. Their color is white or rose.

FOLIAGE The opposite leaves are smooth and toothed along their edges and lance-shaped.

ASSETS Handsome foliage and intriguing flowers at a time when few native plants are in bloom. They are quite easy to grow.

FAULTS The Turtle-heads are not for poor, dry soil.

CULTURE Most wild-flower specialists offer plants of both species of Turtle-head listed here and ship them in spring or fall. Give them rich, humusy soil, acid or near neutral, that will remain reasonably moist, and a partially shaded location. Once established, clumps will probably self-sow; additional plants are acquired, too, by lifting the clumps in spring or fall and dividing them. Plants are easily transplanted from the wild.

SPECIES *Chelone glabra.* White Turtle-head. (Illustrated) Native from Newfoundland and Manitoba to Florida, west to Alabama and Kansas. Height 1 to 4 feet. White flowers, sometimes pink-tinged, in summer.
C. lyoni. Pink Turtle-head. Native to highlands in the Carolinas and Tennessee. Height about 3 feet. Deep rose flowers in fall.

TWIN-FLOWER

TYPE Evergreen sub-shrub
FAMILY *Caprifoliaceae* (Kap-rif-foh-lee-*ay*-see-ee)
GENUS *Linnaea* (Lin-*nee*-uh)
SPECIES *borealis* (bor-ee-*ay*-liss)
ORIGIN Mountains of Alaska, Vancouver and Newfoundland to Maryland and New Jersey, west to Colorado and California

This exquisite little creeper is in the Honeysuckle family, but even in cold Northern woods, where it abounds, it possesses none of the rambunctious habits of so many of its vining relatives. It most nearly resembles in scale and habit a non-relative, Partridgeberry, but is not as easy to establish. This same plant exists in Northern Europe and Asia and was a favorite of Carl Linnaeus, after whom it was named. The American form is often identified by the varietal name, *americana,* and has slightly larger flowers.

USES As a ground cover in deep woods, especially in and among rocks.

HABIT OF GROWTH Creeping semi-woody stems, up to 24 inches long, which form mats.

FLOWERS Pairs of funnel-shaped, fragrant flowers on short, 4-inch, slender stems from the runners. The flowers, which droop, are ⅓ inch long and pink or white; they appear in late spring.

FOLIAGE Small, rounded evergreen leaves, which are opposite.

ASSETS Twin-flower is a gem of its kind, almost a collector's item for advanced gardeners who enjoy growing rare, more difficult-to-please natives.

FAULTS Too dainty to be showy except in its vast patches in nature. Not for hot, dry woodland gardens.

CULTURE The nurseries which offer Twin-flower usually sell it in matlike sods which should be planted in spring or fall in very acid, humus-rich soil which will retain moisture and be cool in summer. In the wild it is often found in bogs and swamps. It needs full shade. Established plants can be divided after flowering.

TWIN-LEAF

TYPE Herbaceous perennial
FAMILY *Berberidaceae* (Ber-ber-id-*day*-see-ee)
GENUS *Jeffersonia* (Jef-er-*soh*-nee-uh)
SPECIES *diphylla* (dye-*fill*-uh)
ORIGIN Ontario, to Iowa and Wisconsin, south to Alabama

Twin-leaf is in the same family as both the May-apple and Blue Cohosh. Its genus, obviously named after Thomas Jefferson, has only two species, the native American Twin-leaf and a blue-flowered species (*J. dubia*) found only in China.

USES Plant in the woodland garden in colonies or as accents among Ferns and other early spring natives. Twin-leaf is also effective in shaded rock gardens or retaining walls and around rocks.

HABIT OF GROWTH Upright leafy clumps from a scaly base, the leaves and flowering stems being separate from each other.

FLOWERS White-petaled flowers, about 1 inch across, are usually higher than the leaves and last about a day. Each flower is produced singly on a naked stalk, 6 inches high, in early spring. The seed capsule which follows the flower opens at the top to release the seeds.

FOLIAGE Each leaf is divided into two segments which are lobed and roughly kidney-shaped. Their formation easily explains the common name, Twin-leaf. At the time of flowering they are only partially developed, but later may be 20 inches high and tower over the seed capsule left by the flower.

ASSETS A very pretty, out-of-the-ordinary wild flower which is available from most native plant specialists. It is easy to establish in a woodland setting.

FAULTS Not suitable for poor, dry soil.

CULTURE Twin-leaf should be planted in early spring or fall in part shade. The soil, either acid or neutral, should be rich and retain moisture during dry spells. When planting, mix quantities of leafmold and/or peat moss, as well as some clay, into the existing soil if it is excessively sandy. Established clumps are readily divided in late fall. Seed capsules can be collected from wild plants or from those growing in the home garden. Seeds, after being removed from their capsules, can be sown in fall in a mixture of peat moss and sand.

TWISTED-STALK

TYPE Herbaceous perennial
FAMILY *Liliaceae* (Lil-ee-*ay*-see-ee)
GENUS *Streptopus* (*Strep*-top-us)
SPECIES *amplexifolius* (am-plex-if-*foh*-lee-us), *roseus* (*roh*-zee-us)
ORIGIN See Species, below

The Twisted-stalk belongs to the Lily family and is a close relative of several other similar-appearing natives, including Fairy Bells, Solomon's-seal and Merrybells.

USES Fine for naturalizing in a woodland area or on a shaded bank or in the foreground of evergreen shrubs.

HABIT OF GROWTH The plants have thickened, creeping rootstocks from which rise strong, branching stems 2 to 3 feet tall.

FLOWERS Little bells about ½ inch long, singly or in pairs, drooping from stems which bend or twist in the middle (giving the plants their common name), all the way up the main stems, opposite the leaf axils. Flowering time is spring. Red berries follow the flowers.

FOLIAGE The pointed leaves are alternate and stemless, prominently veined, and vary from 2 to 5 inches in length.

ASSETS Attractively formed plants worthy of colonizing in rich, shaded situations.

FAULTS The plants will not thrive in poor, dry soil.

CULTURE Plant the Twisted-stalks in shade in humus-rich soil that is acid and remains cool and moist in summer. Propagation is by cutting apart the thickened root. Seeds contained in the red berries can be removed from their pulpy surrounding and planted in a mixture of peat and sand.

SPECIES *Streptopus amplexifolius.* Twisted-stalk. Native to arctic North America, south to New York and Wisconsin and farther south in the mountains. Green-tinged bells on 3-foot arching stalks, the leaves clasping, glaucous underneath and heart-shaped at the base. Leaves may be 5 inches long. Needs rich, moist soil.

S. roseus. Rosy-bells. Native over Eastern North America, including Labrador and Manitoba, south to New Jersey and Wisconsin, farther south in the mountains. A prettier species because of its rose-colored flowers which show up better than those of the preceding species. The stemless but non-clasping leaves may be 4 inches long. The branching, arching stems are 2 feet long. Flowering time is late spring.

VEIN-LEAF HAWKWEED

TYPE Herbaceous perennial
FAMILY *Compositae* (Kom-*poz*-it-ee)
GENUS *Hieracium* (Hye-er-*ray*-see-um)
SPECIES *venosum* (ven-*noh*-sum)
ORIGIN Maine to Georgia and Nebraska

This Hawkweed, a close relative of the Dandelion, is known by a variety of common names, but Vein-leaf Hawkweed seems the most appropriate. Other names include Rattlesnake-weed (for the veining on the leaves), and Poor-robin's-plantain.

USES It colonizes readily in dry woods where it is attractive against large tree trunks, fallen logs or among rocks. It is also suitable for dry soil pockets in informal rock gardens or on rock outcroppings or ledges.

HABIT OF GROWTH From a rosette of nearly ground-hugging leaves rises a wiry stem about a foot tall which bears the flowers.

FLOWERS The golden-yellow ray flowers, about an inch across at the beginning of summer, may measure ½ inch across as the season progresses, and are carried in loose graceful heads at the ends of long, nearly leafless stems. The first—and showiest—flowers appear in late spring but a few are in evidence all summer.

FOLIAGE The leaves, roughly oblong, are about 4 inches long, 1 to 1½ inches wide, and are clustered at the base of the flowering stems to form an irregular rosette. Each leaf is heavily veined in dark red, especially pronounced in spring and early summer.

ASSETS Desirable for naturalizing in dry, open woods or any half-shaded areas where the soil is acid and poor. The foliage rosettes are the chief attraction but the bright yellow flowers waving gracefully above have appeal too.

CULTURE Grow in acid, well-drained soil. The plants are easy to dig when found growing wild; plants in established colonies can be divided in early spring or fall by separating their offshoots. Vein-leaf Hawkweed will self-sow.

VERVAIN

TYPE Herbaceous perennial
FAMILY *Verbenaceae* (Ver-bee-*nay*-see-ee)
GENUS *Verbena* (Ver-*bee*-nuh)
SPECIES *canadensis* (kan-ad-*den*-siss), *hastata* (hass-*tay*-tuh), *stricta* (*strikt*-uh)
ORIGIN See Species, below

These species are in the same genus as the popular garden Verbena, although at first look, it is puzzling to see how the tall Blue Vervains (*Verbena hastata* and *V. stricta*) can be relatives of these sprawling, bright-flowered cultivated plants. On the other hand, *V. canadensis* (illustrated, left) resembles them in both foliage and flower. It is called Clump Verbena or Large-flowered Verbena as well as Midland Vervain. (*Verbena stricta,* illustrated, right, is typical of the tall Vervains.)

USES The Large-flowered Verbena is adaptable to a variety of situations, such as sunny rock gardens, rock outcroppings, or slopes or banks. It can be tucked into soil pockets of informal stone steps or retaining walls. Both species of Blue Vervain need more moisture in their soil and special situations because of their taller habits. In nature they are often found with Ironweed and the Eupatoriums, and when these tall, rugged plants are to be introduced into a naturalistic planting, several clumps of the Blue Vervains should be included.

HABIT OF GROWTH The Large-flowered Verbena makes a sprawling plant, rarely over 2 feet high, its stems ranging from 1 to 2 feet long. Blue Vervain (*V. hastata*) is a very tall plant, its stems from 3 to 4 feet tall or often up to 6 or 7 feet, while *V. stricta* is a slightly shorter counterpart, its stems reaching 30 inches.

FLOWERS Bunched flowers, forming a dome or nearly flat cluster, similar in habit to the Garden Verbena, characterize the Large-flowered Verbena. Each flower is about ⅔ of an inch across and has five petals, at first rosy purple, then fading to rose. Blooming time is late spring and summer. The flowers of the Blue Vervains are tiny, violet-blue and arranged in several short, spear-shaped spikes at the ends of the stems, from midsummer into early fall.

FOLIAGE Toothed or lobed leaves of a coarse texture.

ASSETS Easy-to-grow natives with generous flowering periods, the Large-flowered Verbena producing some flowers in spring and intermittently all summer. The growth habit of the Blue Vervains is intriguing, even though the flowering effect is meager unless several clumps can be massed.

FAULTS The Blue Vervains are considered too weedy for the small wild-flower garden.

CULTURE Grow the Blue Vervains in moderately acid soil enriched with leafmold or peat moss. If the soil is not naturally moist, the plants should be introduced to an area where they can be readily watered during periods of drought. The better the soil, the more lush (and tall) will be the plants. Plants in the wild can be lifted in spring or fall; summer transplanting rarely proves successful. Propagation is easy by division in the spring. Once established, clumps may spread by self-sowing. The Large-flowered Verbena prefers a well-drained soil, acid to neutral, and not too rich. In fact it is well suited to dry, sandy sites, exposed to full sun. Plants can be divided in the spring; nurseries offer plants for spring or fall planting.

SPECIES *Verbena canadensis.* Large-flowered Verbena. Midland Vervain. Rose Verbena. Native from Pennsylvania, south to Florida and west to Illinois and Colorado, Michigan and Minnesota south to Texas.
V. hastata. Blue or Erect Vervain. Widely distributed from Nova Scotia to Florida and west.
V. stricta. Blue Vervain. Native from Ontario to New Mexico. Purple flowers in long dense spikes.

VIOLET

TYPE Herbaceous perennial
FAMILY *Violaceae* (Vye-oh-*lay*-see-ee)
GENUS *Viola* (*Vye*-ol-uh)
SPECIES *adunca* (ad-*unk*-uh), *blanda* (*bland*-uh), *canadensis* (kan-ad-*den*-siss), *conspersa* (kon-*spur*-suh), *eriocarpa* (ehr-ee-oh-*karp*-uh), *palmata* (pal-*may*-tuh), *pedata* (ped-*day*-tuh), *pedatifida* (ped-day-*tif*-fid-uh), *priceana* (pry-see-*ay*-nuh), *pubescens* (pew-*bess*-senz), *rotundifolia* (roh-tun-dif-*foh*-lee-uh)
ORIGIN See Species, below

The Violet is a world-wide favorite, with virtually every region possessing a few, if not many, species. North America is especially rich in Violets. Some inhabit cool, mountainous areas, others, deep Northern woodlands and cold bogs, while others carpet prairies and meadows, dry woods and open sandy banks. Botanists readily accept the close relationship of our garden Pansies and Violas to the usually much smaller-sized native Violets sought for wild-flower plantings, but it seems miraculous to many of us that all belong to the same genus. The eleven species listed at the end of this section are a mere handful of the many species available—either from wild-flower specialists in both plants and seeds or from nearby woods and fields.

USES Probably most of us think of the Violet first as belonging in woodland gardens or as ground-cover plants at the foot of shrubs and trees. In addition to those that should be established in rich, humusy soil in shady areas, there are other species that prefer sun and rather dry, sandy soil, making them good subjects for banks and slopes and rock gardens. Just about everyone has picked Violets at some time and used them in miniature arrangements.

HABIT OF GROWTH Leafy, clumpy plants, some species having a very fibrous root system, others, underground swollen stems. Some are stemless, both leaves and flowers rising directly from the crown, while others, like the Canada Violet, have upright or sprawling stems bearing both leaves and flowers. Some Violets are virtually ground-hugging, others form rounded clumps to 4 or 5 inches or so, but all are generally low-growing.

FLOWERS Despite their common name, Violets are not always violet, purple or blue, although these shades do predominate. There are many white and yellow species. Each flower, a Pansy in miniature, has five petals, often with yellow or white centers. The petals of some are accented by darker veins. Flower size varies from ½ inch to 1½ inches across and except for the Bird's-foot Violet, flowers of the species listed here tend to be non-fertile. Seeds, and usually a good many, are formed later by inconspicuous flowers without petals often hidden by the leaves or nestled close to the ground. Most Violets have fragrance to some degree, that of the Sweet White Violet being especially pronounced. While Violets bloom in spring and early summer, a few bloom again in the fall.

FOLIAGE There is great variation. Some leaves are oval and heart-shaped, others lance-shaped and narrow, and a few have dissected and deeply cut leaves.

ASSETS Dainty flowers, usually in profusion, and attractive foliage. Most Violets are adaptable and easy to grow if reasonable care is taken to provide growing conditions similar to their native environment. Many interesting kinds are listed by wild-flower specialists; others are easy to transplant from the wild.

FAULTS Some Violets spread too fast, and due to their prolific seed production will pop up all over a small garden, becoming a nuisance.

CULTURE Violets are most effective when several plants are set out together to form patches and colonies. The woodland Violets like a humus-rich soil with some shade and generally even moisture, which means the plants may have to be watered in times of drought if the soil does not readily retain moisture. Mix quantities of leafmold and/or peat moss as well as some clay into the planting area if the soil is very sandy. Other Violets survive in poor soil and sun, and, like the Bird's-foot Violet, need such a situation. Most thrive in neutral or acid soil, but the Bird's-foot Violet must have an acid soil. Most Violets self-sow or can be divided. If seed is purchased, it can be sown in the fall, in a mixture of leafmold or peat moss and sand. Young plants will appear in the spring. Specific growing needs are given below.

SPECIES *Viola adunca.* Sand Violet. Native over most of temperate North America from Quebec and Maine westward to the Pacific. Blue to violet flowers, about ½ inch across or slightly less, on short stems 2 to 6 inches long, from mid-spring to early summer. An attractive little species for light shade or sun and rather lean, sandy soil with good drainage.

V. blanda. Sweet White Violet. Native from Quebec to Minnesota, Indiana and Wisconsin, south to Georgia. White flowers, ½ inch across, scented, on 3-inch high plants, in early spring. Prefers some shade and rather rich, moist soil, but adapts to average conditions. Plants may be divided. It will form attractive colonies.

V. canadensis. Canada Violet. Newfoundland to Saskatchewan, south to South Carolina, Alabama, Iowa and Arizona. Leafy, tall species to 1 foot with white flowers with a yellow center, the petals sometimes tinged violet, in spring and early summer. Prefers cool, rich, woodland conditions. Easily divided in the fall.

V. conspersa. American Dog Violet. Native from Quebec to Minnesota, south to Georgia. Lovely pale blue flowers above 6-inch leafy plants from early to late spring. This species, especially generous with its flowers, likes rich, moist soil and some shade. It can be divided.

V. eriocarpa. Smooth Yellow Violet. Nova Scotia to Manitoba, south to Georgia and Texas. Bright yellow flowers, ¾ inch across, on short stems from the leaf axils, in spring and early summer. Height, about 10 inches. Likes light shade, moist woodsy soil. Can be divided.

V. palmata. Early Blue Violet. Massachusetts to Minnesota, south to Florida. Violet-purple flowers, with white veins about 1 inch across, in early spring to early summer, on stems as high as the leaves, about 7 inches. An attractive species for average conditions. Can be divided.

V. pedata. Bird's-foot Violet. Pansy Violet. Showy, large flowers more nearly resembling the garden-cultivated Pansy and *Viola cornuta.* This lovely species has flowers which may be 1½ inches across, the petals lilac-purple or the upper petals deeper colored. The foliage is unique too, the leaves being divided into three or more segments, each segment again deeply cleft or toothed, somewhat resembling a bird's foot, as the common name implies. Native from Massachusetts to Florida, to Minnesota and Louisiana. The Bird's-foot Violet needs acid soil, which should be well drained and lean. It is usually seen in sandy or gravelly soil, in full sun or light shade. The flowers appear in spring and early summer, then the foliage disappears, the rosettes appearing again in the fall. This species is suitable for pockets in rock walls and the rock garden if the soil is acid and gritty; or group it in open, dry woodlands. It self-sows readily. Most wild-flower nurseries offer plants, listing also *V. pedata bicolor,* which has purple-black upper petals.

V. pedatifida. Prairie Violet. Larkspur Violet. Ohio to Arizona. Violet-purple flowers, 1 inch across, above deeply cut leaves. The plants are about 4 inches high. This species somewhat resembles the Bird's-foot Violet but has smaller flowers and leaves a little like those of the Larkspur. Grow it in light shade or sun in neutral or acid soil, rich in humus but well drained.

V. priceana. Confederate Violet. Native from Kentucky to Georgia and Arkansas. Some botanists consider this a form of the common Violet, *V. papilionacea.* It is a handsome, strong species with white flowers strongly veined with blue, very pretty when blooming in mid-spring. Later its large clumps of heart-shaped leaves may be weedy. Nevertheless this fast-spreading species makes an excellent carpeting plant where space is adequate and there is no danger of it crowding out more delicate plants. It thrives in light shade in rich moist soil, but will do quite well under more average conditions.

V. pubescens. Downy Yellow Violet. Nova Scotia to Virginia and Missouri. Similar to the Smooth Yellow Violet, but stems and leaves are covered with down. It likes humus-rich but well-drained soil in some shade.

V. rotundifolia. Round-leaved Yellow Violet. Maine to Ontario, south to Georgia. Small mat-forming species with bright yellow flowers veined in brown in early spring. When this species blooms, its heart-shaped leaves are barely 1 inch wide; by midsummer they may be 2 to 4 inches wide. Prefers cool, rich woodland conditions. Suitable for shaded portions of a rock garden or for colonies in a woodland garden.

VIRGINIA-BLUEBELLS

TYPE Herbaceous perennial
FAMILY *Boraginaceae* (Bor-aj-in-*nay*-see-ee)
GENUS *Mertensia* (Mer-*ten*-see-uh)
SPECIES *virginica* (vir-*jin*-ik-uh)
ORIGIN New York and Ontario, to South Carolina and Alabama, to Minnesota, Kansas and Arkansas

Virginia-bluebells is also known as Virginia-cowslip and is an obvious member of the Borage or Forget-me-not family with its true-blue flowers, usually quite pink in the bud stage. While Virginia-bluebells is the only species given here, it is by no means the beginning or end of the genus, as there are many other similar species native to the Rockies and Pacific Northwest as well as the Himalayas. The Western and Asiatic species are as lovely as the species listed here and, when available, are certainly worth a trial although they are generally considered less adaptable.

USES Grow in clumps as accents among Daffodils, Ferns and early Tulips, or allow the plants to colonize large or small areas in open woodlands. Virginia-bluebells is often included in semi-shady portions of perennial borders with Bleeding-heart (*Dicentra spectabilis*), Leopard's-bane (*Doronicum*) and Primroses, to name a few. Plants are also suited to semi-shaded sections of a rock garden or in groups at the base of a stone wall. In fact, there's hardly a home garden which doesn't have at least one situation where a few plants of this desirable native can't be satisfactorily grown.

HABIT OF GROWTH Upright or slightly bending or arching stems, smooth, to about 2 feet, from a thick tuberous root.

FLOWERS Terminal clusters of funnel-shaped blue flowers, about 1 inch long and pink in the bud. Flowering time is from early to mid-spring.

FOLIAGE Smooth, alternate leaves, oblong in shape, and from 2 to 5 inches long.

ASSETS A choice and adaptable native, deserving of its long popularity because of its lovely blue flower display in the spring.

FAULTS This is one of the wild flowers which lose their top growth soon after the seeds have ripened, usually in early summer, so care must be taken not to dig into its underground parts.

CULTURE Virginia-bluebells needs semi-shade or light shade, and humus-rich soil which retains some moisture. (It is not suitable for poor, bone-dry soil.) The fleshy roots can be planted in early spring, or while dormant in summer and fall, and should be set so they are covered by 4 inches of soil. Add peat moss, leafmold or compost to the planting holes, and some clayey soil if the existing soil is predominantly of sand. Existing plants may self-sow or seed can be purchased. Sow in spring or fall in a mixture of peat moss and sand. Long-established clumps that have become crowded can be lifted after flowering, divided, and replanted.

WILD CALLA

TYPE Herbaceous perennial
FAMILY *Araceae* (Ar-*ray*-see-ee)
GENUS *Calla* (*Kal*-luh)
SPECIES *palustris* (pal-*lust*-riss)
ORIGIN Canada, south to Pennsylvania, Indiana and Minnesota

This water-loving plant, a close relative of the Jack-in-the-pulpit, looks a little like the tropical Calla-lily, which is also in the same family. Another name for it is Water-arum, and the species listed here is the only one in the genus.

USES In bogs and swamps and along margins of ponds and pools.

HABIT OF GROWTH The Wild Calla has a horizontal, fleshy rhizome, really a stem, which creeps. Each rootstock produces several long-stemmed leaves and a single flower up to a foot or less in height.

FLOWERS Clusters of small green flowers which form club-shaped spikes in early summer, followed by red berries. The showy part of the plant is the white bract or spathe, irregularly shaped and nearly flat, which backs the flower spike.

FOLIAGE Prominently veined, heart-shaped leaves on fleshy petioles.

ASSETS An interesting accent for pools and bogs.

FAULTS Suitable only for water gardens or very wet soil.

CULTURE The Wild Calla needs rich, cool, acid soil that remains wet or moist, and full sun or light shade. It is fine in shallow pools or at the edges of ponds, where its creeping rhizomes should be barely sunk in the mud with a few inches of water covering them. Plants will do quite well in bogs, man-made or natural, where the soil never completely dries out. The Wild Calla produces a new rootstock each year, after flowering and fruiting are completed. Plants may be increased by separating the rhizomes in spring, allowing a bud or "eye" to each rhizome division. Seeds, when available, can be separated from the surrounding pulp and sown in a flat first lined with plastic, then filled with a peat-and-sand mixture or, better yet, natural bog soil. The plastic will help the soil mixture retain constant moisture. The seedlings can remain in the flat until they are large enough to handle and be transplanted into permanent locations.

WILD GERANIUM

TYPE Herbaceous perennial
FAMILY *Geraniaceae* (Jer-ray-nee-*ay*-see-ee)
GENUS *Geranium* (Jer-*ray*-nee-um)
SPECIES *maculatum* (mak-yew-*lay*-tum)
ORIGIN Ontario, Maine and Manitoba, south to Georgia and Alabama and west to Kansas

Wild geranium, also known as Crane's-bill, with its pretty magenta or lavender-pink flowers, is a familiar sight in late spring along country roads, in stony soil, either dry or moist, or in richer woodlands. It grows with such abundance that the temptation to pick bouquets of it cannot be resisted, but the flowers always immediately wilt and rarely revive. (Herb Robert, *Geranium robertianum,* is a rather weedy annual or biennial species which closely resembles this species, although both its leaves and reddish purple flowers are smaller. Its leaves, which smell like musk when crushed, have given it the alternate common name of Musk Geranium.)

USES An excellent colonizer, especially effective among Ferns and other spring flowers in shade. It is useful in shaded parts of a large rock garden, and quickly fills in rock-studded banks.

HABIT OF GROWTH Several erect, hair-covered stems, 10 to 24 inches high, grow from a swollen knobby root.

FLOWERS Terminal clusters of pretty, rose-purple flowers, 1 to 1½ inches across, with five petals which are veined a deeper color, in mid-spring to early summer. A rare white-flowered form is occasionally offered. The fruit capsule that follows is shaped like a crane's bill.

FOLIAGE Attractive, deeply cut and notched leaves on long petioles.

ASSETS The abundant flowers over a long period, the plant's readiness to spread into colonies, its tolerance of either rather dry or moist woodland soil. It is listed by all wild-flower specialists.

FAULTS In a small wild-flower garden of rich soil, the Wild Geranium may eventually take over the area if plants are not occasionally thinned out.

CULTURE While this friendly native prefers rich, humusy soil that retains moisture, it does quite well in average soil. It needs at least partial shade. Plants will self-sow freely—sometimes too freely! Clumps are readily divided in spring or fall.

WILD-GINGER

TYPE Herbaceous or evergreen perennial
FAMILY *Aristolochiaceae* (Ar-rist-oh-loh-kee-*ay*-see-ee)
GENUS *Asarum* (*Ass*-uh-rum)
SPECIES *canadense* (kan-ad-*den*-see), *shuttleworthi* (*shuttle*-worth-eye)
ORIGIN See Species, below

Wild-ginger is intriguing but it is apt to be overlooked because its odd, short-stemmed flowers are usually hidden by its own leaves as well as by partially decaying debris when they appear in early spring. The rhizome of Wild-ginger has a pleasant flavor and, like the root of the true Ginger of commerce, which is no relation, may be candied. There are at least two similar species of Wild-ginger native to the Northwest and mountains of California.

USES An excellent ground cover in shade or woods, or as a foreground planting to Rhododendrons and other evergreen shrubs. Wild-ginger will cover a shaded bank or slope, will grow among rocks and boulders, in rich soil pockets of naturalistic rock gardens or in retaining walls.

HABIT OF GROWTH Low-growing plants which spread by their creeping, underground rhizomes.

FLOWERS Odd, three-lobed, tubular flowers, rich maroon or brownish, about one inch across, borne singly on very short stems, sometimes before the leaves have fully unfolded. They are very long-lasting, sometimes barely protrude from the soil. One way partly to solve the problem of seeing the flowers is to establish Wild-ginger on steep banks or at the top of rock walls. When viewed from below, the flowers show up quite well.

FOLIAGE Attractive, heart-shaped leaves in pairs and on long petioles from the rhizome. The petioles vary in length from 4 or 5 inches to nearly a foot.

ASSETS A useful plant for shade, which can form attractive colonies.

FAULTS Not for sun and not suited to poor, dry woodland.

CULTURE Wild-ginger needs the rich, humusy soil suited to most Ferns, both true and False Solomon's-seal and May-apple. The soil can be acid or neutral but should be capable of retaining moisture during periods of drought. It will tolerate sun in spring, but through the summer needs deep or part shade. Plants are easy to collect in the spring or fall.

SPECIES *Asarum canadense.* Wild-ginger. Indian-ginger. (Illustrated) Native from New Brunswick to Manitoba, south to North Carolina and west to Missouri and Kansas. Brown or velvety maroon flowers, 1 inch across, from early to late spring. The paired leaves, often dark brown or red in the spring, are soft and hairy and may reach 7 inches across.

A. shuttleworthi. Mottled Wild-ginger. Native from Virginia to Georgia and Alabama. This attractive species has proved to be hardy beyond its more southerly range. Its paired leaves, thick-textured and mottled with white, are evergreen and aromatic and about 3 inches wide.

WILD-PETUNIA

TYPE Herbaceous perennial
FAMILY *Acanthaceae* (Ak-anth-*ay*-see-ee)
GENUS *Ruellia* (Rew-*ell*-ee-uh)
SPECIES *ciliosa* (sil-ee-*oh*-suh)
ORIGIN New Jersey to Nebraska, south to Florida and Texas

This widely distributed native plant has flowers which somewhat resemble those of the true Petunia, although they do not even share the same family. There are several other similar species of Ruellia common to this country and South America, but this species is the one most often offered by specialists in either seed or plant form. It is also called Fringe-leaf Ruellia.

USES For colonizing in part shade or sun in open patches in woodland, along informal paths, on slopes, in rock gardens or against walls. Good companions include any of the early Goldenrods and the Golden-aster.

HABIT OF GROWTH Strong, fibrous-rooted clumps with several hairy stems to about 12 inches.

FLOWERS The tubular flowers, which grow from the ends of the stems and from the leaf axils, are about 2 inches long, and of a rich blue-lavender color. Although each flower lasts only a day, there is an almost continual supply of new buds. Flowering time is from midsummer, often earlier, into fall.

FOLIAGE Opposite leaves, with a fringe of hairs along their edges, and about 3 inches long.

ASSETS Tolerance of dry, even sandy, soil and its midsummer flowering habit make this a useful native for dry open woods or sunny areas where the soil is far from ideal for more demanding plant material.

CULTURE Slightly acid or near neutral soil that can be sandy will suit the Wild-petunia. It will thrive in sun or part shade and will self-sow. Seeds, usually available from specialists, can be sown in spring. Plants are readily divided in early spring.

WINE-POPPY

TYPE Herbaceous perennial
FAMILY *Malvaceae* (Mal-*vay*-see-ee)
GENUS *Callirhoë* (Kal-*lihr*-eh-ee)
SPECIES *involucrata* (in-vol-yew-*kray*-tuh)
ORIGIN Missouri to Wyoming, south to Texas and New Mexico

Another name for Wine-poppy is Wine-cups, and both names perfectly describe the flowers. Actually this native is not in the Poppy family and is more nearly related to the Swamp Mallow and garden-cultivated Hollyhock.

USES Wine-poppy is suitable for large rock gardens and walls, slopes and banks and any warm, dry areas where its semi-procumbent habit will be effective.

HABIT OF GROWTH Creeping or trailing stems from a thick taproot.

FLOWERS Five-petaled flowers, cupped like a single Rose and about 2½ inches across, in fair profusion from late spring to late summer. Their color is a rich cherry-red which can be very conspicuous.

FOLIAGE Alternate, lobed and deeply cut.

ASSETS The flowers are very showy and appear over a long period, the plants rarely being without bloom during the summer. The plants are drought-resistant and thrive in poor soil.

FAULTS As with all taprooted plants, the Wine-poppy resents being disturbed once it has become established. It is too weedy for restricted rock gardens.

CULTURE Most general-purpose nurseries as well as some nurseries specializing in native plants offer plants of the Wine-poppy for early spring or fall planting. The long, thick root should be set upright with its crown level with the surface in well-drained soil that is predominantly sandy or gritty in texture. Mix some leafmold or peat moss in the planting hole if the soil is excessively sandy. Established plantings of Wine-poppy will increase by self-sowing, and may become a nuisance.

WINTERGREEN

TYPE Evergreen sub-shrub
FAMILY *Ericaceae* (Ehr-ik-*ay*-see-ee)
GENUS *Gaultheria* (Gol-*theer*-ee-uh)
SPECIES *procumbens* (proh-*kumm*-benz)
ORIGIN Newfoundland to Manitoba and Georgia

For many, a happy memory of woods roaming was finding a patch of Wintergreen, half immersed in forest litter, and tasting the pulpy berries and leaves with their Wintergreen flavor. (This species and probably other non-native species were once the sole source of this flavoring.) Wintergreen, also known as Checkerberry or Teaberry, is yet another member of the Heath family (*Ericaceae*), which includes such outwardly dissimilar genera as Rhododendron, Bearberry and Trailing-arbutus. There are many other species of *Gaultheria,* mostly natives of Asia, which have been introduced to this country and make excellent ground covers in shade.

USES In its native haunts in cool, rich woodland, Wintergreen can make a lush ground cover. When introduced to the home wild-flower garden, it may cover patches of ground slowly. It is certainly suited to planting in rich soil pockets in shaded rockeries or at the base of partly submerged rocks, and seems to enjoy the company of Conifers, Oaks and broad-leaved evergreens of its own family.

HABIT OF GROWTH Low-growing, up to 6 inches, spreading by underground stems which are semi-woody.

FLOWERS The nodding flowers are white, ¼ inch long, bell-shaped and similar to those of Blueberry and Bearberry, its very near relatives. They appear irregularly in summer and are followed by bright red berries which may persist for several months.

FOLIAGE Aromatic, alternate and evergreen leaves, oval-shaped and from ½ to 2 inches long. Their winter color is reddish or bronze.

ASSETS For some, this is a choice native, ever intriguing because of its delicate flowers and flavored berries. It is ideal for introducing to a cool, rich, damp woodland.

FAULTS Rather straggly and sparse in its growth habit in dry, poor woods. Its production of flowers and fruit can be sparse, whole colonies seeming to produce none. Wintergreen must have acid soil.

CULTURE Wintergreen grows to perfection in acid soil, rich in leafmold. It also prefers shade but if the soil is moist enough, will tolerate some light or sun. Clumps can be fairly successfully lifted in spring or fall and should be planted in well-prepared soil full of peat moss and/or leafmold. Regular watering is necessary until the plants become established. In late spring and summer, non-rooted stems can be put in flats of peat moss and sharp sand and covered with a tent of polyethylene after being well watered. They will root in a few weeks. Most wild-flower specialists offer plants.

WOOD-BETONY

TYPE Herbaceous perennial
FAMILY *Scrophulariaceae* (Skrof-yew-lay-ree-*ay*-see-ee)
GENUS *Pedicularis* (Ped-ik-yew-*lay*-riss)
SPECIES *canadensis* (kan-ad-*den*-siss)
ORIGIN Nova Scotia and Manitoba to Florida and Mexico

Sooner or later, nearly everyone who wanders through spring woodlands will have encountered the Wood-betony. Though not the showiest spring flower, a large clump of Wood-betony is startling, possibly because of the flowers' out-of-season color combination of red and yellow which superficially makes one think of a Chrysanthemum. Another name for the genus is Lousewort. There are several other species native to North America, the most spectacular one being found in the West.

USES For accent in open woodlands, especially along grassy paths, and in naturalistic rock gardens.

HABIT OF GROWTH Erect or slightly sprawly clumps to about 12 inches.

FLOWERS Tubular red and yellow flowers, about ¾ of an inch long, in loose heads or short spikes, from spring to early summer.

FOLIAGE The alternate leaves are very attractive, varying in length from 3 to 6 inches. They are deeply cut and Fernlike and often show a reddish cast over their green coloring.

ASSETS Interesting plants, effective for their long-blooming flowers and distinctive foliage.

FAULTS Plants, either collected or purchased from nurseries, are often hard to establish in the home wild-flower garden.

CULTURE Those who find patches of the plant nearby will have success transplanting clumps to their own place if they dig them in large, deep sods and provide a similar growing site. The Wood-betony may be partially parasitic on roots of other plants. In its native habitat, Wood-betony grows like a weed, but often sulks when transplanted. The plants are usually found in open woodlands and sometimes in meadows in moderately rich soil that is acid and retains moisture.

YELLOW STAR-GRASS

TYPE Bulbous perennial
FAMILY *Amaryllidaceae* (Am-are-ril-lid-*day*-see-ee)
GENUS *Hypoxis* (Hye-*pox*-iss)
SPECIES *hirsuta* (herss-*yew*-tah)
ORIGIN Maine to Ontario, Kansas, Florida and Texas

This is a most satisfying little plant for the small wild-flower collection. It is one of the few natives of the Amaryllis family, which also contains Daffodils and Snowdrops.

USES The Yellow Star-grass is most effective when established in colonies along paths in the open woodland garden or in a sunny rock garden. Plants show up especially well when nestled against large stones, or set out on slopes or banks, with or without rocks. It is a good companion to Blue-eyed-grass.

HABIT OF GROWTH Clumps of Grasslike foliage, longer than the flowering stems, rising from a corm or bulbous rootstock.

FLOWERS Lemon-yellow, six-petaled, star-shaped flowers, about ½ inch wide, carried in umbels of five to seven flowers at the end of 6- to 12-inch stems, lower than the leaves. Both stems and flower buds are covered with tiny hairs. The blooming period is a long one, the first flowers appearing in late spring, but after a short rest, some flowers appear throughout the summer.

FOLIAGE The Grassy or ribbonlike leaves are attractive. They are about ¼ of an inch wide, 2 feet long, and also covered with white hairs.

ASSETS An easy-to-grow wild flower of great charm. It will settle down in a variety of situations.

FAULTS Bound to be overlooked by those who value size and flamboyance first.

CULTURE Grow in sun or light shade, in dry, acid soil. Plants, available from most wild-flower nurseries, can be set out in spring and fall. Established plants can be readily divided by removing offsets. Add peat moss or leafmold to very sandy soil before setting out clumps.

SOURCES OF WILD FLOWERS

ALPENGLOW GARDENS
13328 Trans-Canada Highway
North Surrey P. O.
New Westminster, British Columbia
(Dwarf Shrubs, Rockery Plants and Some Wild Flowers)

AMERICAN PERENNIAL GARDENS
6975 Dover Street
Garden City, Michigan
(Wild Flowers, Seeds)

AMERICAN ROCK GARDEN SOCIETY
Secretary: Lawrence Hochheimer
Ridge Farms Road
Norwalk, Connecticut
(Membership in society costs $5.00. The interesting bulletin always contains some information on Native Plants and there is a seed exchange which includes many Wild Flowers)

ARTHUR EAMES ALLGROVE
North Wilmington, Massachusetts
(Wild Flowers, Ferns, Insectivorous Plants, Partridge-berry Bowls and Supplies Related to Terrariums. Catalog costs 25 cents.)

BALDWIN'S SEED CO. OF ALASKA
Anchorage, Alaska
(Seeds of Wild Flowers)

CLAUDE A. BARR
Prairie Gem Ranch
Smithwick, South Dakota
(Wild Flowers and Seeds)

CAROLINA BIOLOGICAL SUPPLY COMPANY
Burlington, North Carolina
(Insectivorous Plants and Terrarium Supplies)

FERNDALE NURSERY AND GREENHOUSES
Askov, Minnesota
(Wild Flowers and Ferns)

GARDENS OF THE BLUE RIDGE
Ashford, North Carolina
(Wild Flowers, Ferns, Native Shrubs and Evergreens)

GARDENSIDE NURSERIES, INC.
Shelburne, Vermont
(Wild Flowers, Ferns, etc.)

GREEN BUSH GARDENS
Charlotte, Vermont
(Wild Flowers, Ferns, Trees for Woodlands)

RUTH HARDY'S WILD FLOWER NURSERY
Falls Village, Connecticut
(Wild Flowers and Ferns)

LESLIE'S WILD FLOWER NURSERY
30 Summer Street
Methuen, Massachusetts 01844
(Wild Flowers, Seeds and Ferns)

LOUNSBERRY GARDENS
Oakford, Illinois
(Wild Flowers and Ferns)

WALTER MARX GARDENS
Boring, Oregon
(Trilliums, Trout-lilies and Bulbs for Woodlands)

MAYFAIR NURSERIES
R.D. 2, Nichols, New York
(Bearberry, Dwarf Shrubs, etc.)

MOUNTAIN VALLEY NURSERY
Valley Road
Jefferson, New Hampshire
(Wild Flowers and Rock Garden Plants)

L. L. OLDS SEED COMPANY
722 Williamson Street
Madison, Wisconsin
(Wild-flower Plants)

ORCHID GARDENS
Route One, Box 290
Grand Rapids, Minnesota 55744
(Wild Flowers, Ferns, Native Trees and Shrubs, etc.)

PEARCE SEEDS AND PLANTS
Moorestown, New Jersey 08057
(Wild Flowers and Seeds)

PUTNEY NURSERY, INC.
Putney, Vermont
(Wild Flowers, Native Trees, Shrubs and Evergreens)

CLYDE ROBIN
P. O. Box 2091
Castro Valley, California
(Seeds of Wild Flowers, Trees and Shrubs. Catalog costs 50 cents)

FRANK H. ROSE
1020 Poplar Street
Missoula, Montana
(Seeds of Wild Flowers of West)

HARRY E. SAIER
Dimondale, Michigan
(Seeds. Catalog costs 50 cents)

SAVAGE GARDENS
Box 163
McMinnville 3, Tennessee
(Wild Flowers)

SKY-CLEFT GARDENS
Camp Street Ext.
Barre, Vermont
(Wild Flowers, Ferns and Rockery Plants)

SKY HOOK FARM
Johnson, Vermont
(Primroses to grow with Woodland Plants)

STONECROP NURSERIES, INC.
R.F.D. 1
Cold Spring, Putnam County
New York
(Wild Flowers and Rockery Plants)

SUNNYBROOK FARMS NURSERY
9448 Mayfield Road
Chesterland, Ohio
(Wild Flowers, Herbs)

THREE LAURELS
Marshall, North Carolina
(Wild Flowers, Ferns, Native Shrubs, Trees and Evergreens)

THURMAN'S GARDENS
Route 2, Box 259
Spokane, Washington
(Wild Flowers, Seeds, Alpine and Rockery Plants. Catalog costs 50 cents)

VICK'S WILDGARDENS, INC.
Box 115, Gladwyne, Pennsylvania
(Wild Flowers and Ferns)

WAKE ROBIN FARM
Route 1, Box 33
Home, Pennsylvania
(Wild Flowers and Ferns)

THOMAS M. WOOD
Constantia, New York
(Wild Flowers and Ferns)

WOODLAND ACRES NURSERY
Crivitz, Wisconsin
(Wild Flowers and Ferns)

INDEX

Photographs are indicated by page numbers in **boldface**

Actaea
  alba, 14
  arguta, 14
  pachypoda, 12, 14
  rubra, 12, 14
Actinomeris alternifolia, 34
Adder's-tongue, 254
Alum-root, 1, 2
American Penstemon Society, 22
American Rock Garden Society, 296
Anemone, 4
  Broad-leaf, 6
  Canada, 6
  Meadow, 6
  Round-leaved, 6
Anemone
  canadensis, 4, 6
  caroliniana, 4, 6
  patens, 4, 6
  pulsatilla, 6
  quinquefolia, 4, 6
  virginiana, 4, 7
Anemonella thalictroides, 5, 216
Aquilegia
  caerulea, 58, 60
  canadensis, 58, 60
  chrysantha, 58, 60
  jonesi, 58, 60
  longissima, 58, 60
  skinneri, 58, 60
Arctostaphylos uva-ursi, 18
Arisaema
  dracontium, 147
  triphyllum, 146
Asarum
  canadense, 282, 284
  shuttleworthi, 282, 284
Asclepias
  incarnata, 53
  purpurascens, 53
  tuberosa, 52
Aster, 8–11
  Blue Wood, 10
  Calico, 10
  Heath, 10
  Large-leaved, 11
  New England, 11
  New York, 11
  White, 11
  White Upland, 11
  White Wood, 10
Aster
  cordifolius, 8, 10
  divaricatus, 8, 10
  ericoides, 8, 10
  kumleini, 8, 10
  laevis, 8, 10
  lateriflorus, 8, 10
  macrophyllus, 8, 11
  novae-angliae, 8, 11
  novi-belgi, 8, 11
  ptarmicoides, 8, 11
  umbellatus, 8, 11

Baneberry, 12
  Red, 14
  White, 14
Baur, Robert C., 194, 239
Beach-heather, 137

Bead-lily, 15
  Blue, 17
  Speckled, 17
  White, 17
  Yellow, 17
Bearberry, 18, **88**
Beard-tongue, 21
Bee-balm, 26
Beetleweed, 101
Bergamot, 25
  Lavender, 27
Birthroot, 251
Bishop's-cap, 28
Bitter-root, 30
Black Cohosh, 50
Black-eyed Susan, 33
Blazing Star, 82, 104
Bleeding-heart, 71
  Fringed, 71
  Plumy, 72
  Western, 73
  Wild, 72
Bloodroot, 36, **243**
Blooming Sally, 93
Blue Cohosh, 41
Bluebeads, 16, 17
Bluebell, 39
Bluebells-of-Scotland, 40
Blue-eyed-grass, 43
Bluets, 46, **90**
  Creeping, 48
  Mountain, 48
  Purple, 48
Bog Garden, making artificial, 16, **244–245**
Boneset, 75, 76
Bradley, Howard S., 166
Brown-eyed Susan, 34
Bugbane, 49
  American, 51
  Cohosh, 51
Bunchberry, viii, 16
Buttercup, Swamp, 247
Butterfly-weed, 52

Cactus
  Brittle, 212
  Prickly-pear, 212
Calla, Wild, 278
Calla palustris, 278
Callirhoë involucrata, 287
Calopogon pulchellus, 130
Caltha palustris, 171
Campanula
  petiolata, 40
  rotundifolia, 39
Canada Mayflower, 83
Candy-root, 186
Candy-weed, 186
Cardinal-flower, 166, 167
Catchfly, 55
  Pink, 56
  Red, 56
Caulophyllum thalictroides, 41
Chamaelirium luteum, 81
Chamise-lily, 255
Checkerberry, 290
Chelone
  glabra, 256, 257
  lyoni, 256, 257
Chimaphila
  maculata, 203, 204, 205
  umbellata, 203, 205
Chiogenes hispidula, 64
Chrysanthemum leucanthemum, 34
Chrysogonum virginianum, 123
Chrysopsis
  falcata, 113
  mariana, 113
  villosa, 113
Cimicifuga
  americana, 49
  racemosa, 49
Claytonia

caroliniana, 238, 239
lanceolata, 240
rosea, 240
virginica, 238, 240
Clintonia, Red, 16
Clintonia
andrewsiana, 16
borealis, 15
umbellata, 15
uniflora, 16
Cloth-of-gold, 137
Cohosh, 13, 14
Black, 50
Blue, 41
Columbine, 58–60
American, 60
Colorado, 60
Eastern, 60
Golden-spurred, 60
Wild, 60
Coneflower, 34
Green-headed, 34
Thin-leaved, 34
Coptis
groenlandica, 125
trifolia, 126
Coral-bells, 1
Corn-lily, 17
Corydalis, 61
Golden, 62
Pink, 62
Pale, 63
Corydalis
aurea, 61
bulbosa, 62
cava, 62
sempervirens, 61
Cowslip, American, 172
Crane's-bill, 281
Creeping Snowberry, 64
Crinkleroot, 66
Three-leaved, 67
Cucumber-root, Indian, 138
Culver's-root, 68
Cypripedium
acaule, 151, 152, 153, 154
calceolus, 151, 152, 154
candidum, 151, 155
pubescens, 154
reginae, 151, 153, 155
spectabile, 151, 155

Daisy, 34, 114
Field, 34
Darlingtonia californica, 207
Day-lily, 161
Deer-grass, 178
Dentaria
diphylla, 66
laciniata, 66
Devil's-bit, 82
Dicentra, 70
Dicentra
canadensis, 70
cucullaria, 70
eximia, 70
formosa, 70
spectabilis, 71
Dionaea muscipula, 207
Disporum lanuginosum, 78
Dodecatheon
cusicki, 224, 225
meadia, 224, 225
pulchellum, 224, 226
Dogtooth-violet, 254
Doll's-eyes, 13
Donaldson, Renée du Pont, 154
Dragon-root, 147
Drosera, 207
Dutchman's-breeches, 72

Epigaea repens. See Trailing-arbutus
Epilobium angustifolium, 92
Epipactis, 214

Eupatorium, 74
  Hyssop-leaved, **156**
Eupatorium
  coelestinum, 74
  dubium, 74
  hyssopifolium, 74
  maculatum, 74
  perfoliatum, 74
  purpureum, 174

Fairy Bells, 78
Fairy-candles, 50
  Appalachian, 51
Fairywand, 81
False-heather, 137
False Lily-of-the-valley, 83, **90**
False-miterwort, 95
False Solomon's-seal, 85
  Starry, 87
False Spikenard, 87
Fawn-lily, 254, 255
  Midland, 255
  White, 255
Fireweed, 92
Flag, Blue, 141, 142, 143
Flowering-wintergreen, 187
Foam-flower, 94, **158**
Fringed Orchid, 96
Fumitory, 62
Fume-root, 62

Galax, 100
Galax aphylla, 100
Galaxy, 101
Gaultheria procumbens, 65, 289
Gayfeather, 103
  Cat-tail, 105
  Prairie, 105
  Spike, 105
Gaywings, 186, 187
Gentian
  Bottle, 108
  Closed, 108
  Fringed, 109
  Soapwort, 109
Gentiana
  andrewsi, 106
  crinita, 106
  decora, 106
  saponaria, 106
Geranium, Wild, 280
Geranium
  maculatum, 280
  robertianum, 281
Ginseng, 110
  Dwarf, 112
  True, 112
Glacier-lily, 255
Gloriosa-daisy, 34
Golden-aster, 113
  Broad-leaf, 115
  Maryland, 115
  Midland, 115
Golden Glow, 34
Golden Groundsel, 117
Golden Ragwort, 116
Goldenrod, 118
  Alpine, 120
  Anise-scented, 120
  Canada, 120
  Fragrant, 120
  Gray, 120
  Hoary, 120
  Velvety, 120
Goldenseal, 121
Golden-star, 123, **156**
Gold-heather, 137
  Woolly, 137
Goldthread, 125
Goodyera pubescens, 213
Grass-of-Parnassus, 127
Great Willow Herb, 93
Greek-valerian, 150
Green-dragon, 147
Ground-raspberry, 122
Groundsel, Golden, 117

Habenaria
ciliaris, 96
fimbriata, 96
psycodes, 96
Harebell, 40
Western, 40
Hawkweed, Vein-leaf, **246,** 265
Heartleaf-lily, 84
Helenium autumnale, 34
Helmet-flower, 230
Hemerocallis, 161
Hepatica, 132
Round-lobed, 134
Sharp-lobed, 134
Hepatica
acutiloba, 132
americana, 132
Herb Robert, 281
Heuchera
americana, 1
sanguinea, 2
Hieracium venosum, 265
Horse-mint, 26
Houstonia
caerulea, 46
purpurea, 46
serpyllifolia, 46
Hudsonia, **88,** 135
Hudsonia
ericoides, 135
tomentosa, 135
Hydrastis canadensis, 121
Hypoxis hirsuta, 294

Indian Cucumber-root, 138
Indian-ginger, 284
Indian Pipe, 204
Iris, 140–141
Copper, 142
Crested, 142
Vernal, 142
Iris
cristata, 140
fulva, 140
missouriensis, 140
pseudacorus, 141
verna, 140
versicolor, 140
Ironweed, 144

Jack-in-the-pulpit, 146
Jacob's-ladder, 149
Jeffersonia
diphylla, 260, 261
dubia, 261
Joe-Pye Weed, 75, 76, 77
Spotted, 76

King-cup, 172
Kinnikinnick, 19

Lady-slipper, 151–155
Pink, 153, 154, **159**
Showy, 153
White, 155
Yellow Slipper, 152, 154
Leopard-lily, 254
Lewisia, 31
Lewisia
rediviva, 30
tweedyi, 30
Liatris
ligulistylis, 103, 105
pycnostachya, 103, 105
scariosa, 103, 105
spicata, 103, 105
Lilium
canadense, 160, 161, 162
grayi, 160, 163
michauxi, 160, 163
michiganense, 160, 163
pardalinum, 160, 163
philadelphicum, 160, 161, 164
superbum, 160, 164
tigrinum, 161
washingtonianum, 160, 164

Lily, 160–164
  American Turk's-cap, 164
  Canada, 161, 162
  Carolina, 163
  Gray's, 163
  Leopard, 163
  Meadow, 162
  Midland, 163
  Mt. Hood, 164
  Orangecup, 164
  Roan, 163
  Shasta, 164
  Tiger, 161, 162
  Wood, 161, 164
Lily-of-the-valley, False, 83
Linnaea borealis, 258
Liverleaf, 133, 134
Lobelia, 165
  Cardinal, 167
  Great Blue, 166, 167
Lobelia
  cardinalis, 165, 166, 167
  siphilitica, 165, 166, 167
Lousewort, 293
Lupine, 168
Lupinus
  arboreus, 169
  perennis, 168, 169
  polyphyllus, 169
  subcarnosus, 169
  texensis, 169

Maianthemum canadense, 83
Mandrake, 174
Manzanita, 19
Marsh-marigold, 171
May-apple, **89,** 173
May-blob, 172
Mayflower, Canada, 84
May-star, 242
Meadow-beauty, 176, 178
  Maryland, 177
Meadow-rue, 179–181
  Early, 181
  Fall, 181
  Purple, 181
  Tall, 181
Medeola virginiana, 138
Merrybells, 182
  Great, 183
  Smooth, 184
Mertensia virginica, 275
Milkwort, 185–187
  Fringed, 186, 187
  Marsh, 186
  Racemed, 186, 187
Mimulus
  alatus, 189
  ringens, 188
Mist-flower, 75, 76
Mitchella repens, 193
Mitella diphylla, 28
Miterwort, 29
Moccasin-flower, 152, 154
Monarda
  didyma, 25
  fistulosa, 25
Monkey-flower, 188
Monotropa
  hypopitys, 204
  uniflora, 204
Moss-pink, 202
Mountain-pink, 202
Mountain-rattlebox, 50

Oconee-bells, 190
Opuntia
  compressa, 210, 212
  fragilis, 210, 212
  humifusa, 212
  polyacantha, 210, 212
Orange-root, 122
Orchid
  Downy Rattlesnake, 214
  Fringed, 96
  Grass-pink, 130

Orange Fringed, 98
Purple Fringed, 98
Rein, 97
Small Purple Fringed, 99
Yellow Fringed, 98, **245**
Yellow Slipper, 152
Orchis, Showy, 227
Orchis
rotundifolia, 228
spectabilis, 227
Oswego-tea, 27

Panax
quinquefolium, 110
trifolium, 110
Parnassia
caroliniana, 127
fimbriata, 127
glauca, 129
palustris, 127
Partridge-berry, 193
Partridge-pea, 196
Pasque-flower, 6
Pea, Large-flowered Sensitive, 197
Pedicularis canadensis, 292
Penstemon
angustifolius, 21
cobaea, 21
digitalis, 21
grandiflorus, 21
hirsutus, 21
Pepper-root, 67
Perennial-ageratum, 76
Phlox, 198–202
Blue, 200
Creeping, 201
Downy, 200
Moss, 200, 202
Sand, 200
Sand-hill, 201
Wild Pink, 201
Phlox
andicola, 198, 199, 200
bifida, 198, 199, 200
divaricata, 198, 199, 200
maculata, 198, 200, 201
ovata, 198, 200
paniculata, 199
pilosa, 198, 200, 201
rigida, 198, 199, 201
stolonifera, 198, 199, 201
subulata, 198, 199, 202
Pine Sap, 204
Pink
Fire, 56
Mountain, *see* Mountain-pink
Wild, 57
Pipsissewa, 203
Green, 205
Mottled, 203, 205
Striped, 203
Pitcher-plant, 206, 209
California, 207
Trumpet, 208
Pleurisy-root, 53
Polemonium
reptans, 150
richardsoni, 150
Polygala
Fringed, 187
Racemed, 187
Polygala
cruciata, 186
paucifolia, 185, 187
polygama, 185, 187
sanguinea, 186
senega, 185, 186, 187
Polygonatum
biflorum, 235, 237
canaliculatum, 237
commutatum, 235, 237
giganteum, 237
Pools, as settings for native plants, **158**
Poor-robin's-plantain, 266

Poverty-weed, 137
Prairie-smoke, 6
Prickly-pear, 210, 212
Prince's-pine, 204, 205
Pyrola, 222
Pyrola
  elliptica, 221, 223
  rotundifolia, 221, 223

Quaker Ladies, 48, **90**

Ragwort, Golden, 116
Ranunculus septentrionalis, 247
Rattlesnake-plantain, 213, **245**
  Downy, 214
Rattlesnake-weed, 266
Rhexia
  mariana, 176, 177
  virginica, 176, 178
Rock gardens, for native plants, **157**
Rosy-bells, 263
Rudbeckia
  hirta, 33
  laciniata, 34
  triloba, 34
Rue-anemone, 5, 216
Ruellia, Fringe-leaf, 286
Ruellia ciliosa, 285

Sanguinaria canadensis, 36
Sarracenia
  flava, 206, 208
  purpurea, 206, 207, 209
Saxifraga
  leucanthemifolia, 219
  michauxi, 218, 219
  virginiensis, 218, 219
Saxifrage
  Early, 219
  Michaux, 219
  Rock, 219
Scutellaria
  incana, 229, 230
  integrifolia, 229, 231
  ovata, 229, 230, 231
  serrata, 229, 231
Senna, Wild, 197
Shinleaf, 221, 223
  Round-leaved, 223
Shooting-star, 224
Shortia galacifolia, 190
Showy Orchis, 227
Silene
  caroliniana, 55
  pensylvanica, 57
  virginica, 55
Sisyrinchium angustifolium, 43
Skullcap, 229, 231
  Downy, 230
  Hyssop, 231
  Larger, 231
  Showy, 231
Skunk-cabbage, 232
Smilacina
  racemosa, 85
  stellata, 85
Snakeroot
  Black, 51
  Large Button, 105
  Seneca, 186
Sneezeweed, 34
Snowberry, Creeping, 64
Solidago
  canadensis, 118
  cutleri, 118
  mollis, 118
  nemoralis, 118
  odora, 118
Solomon's-plume, 86
Solomon's-seal, 85, 235
  True, 237
  Giant, 237
  Great, 237
Spikenard, False, 86

Spring Beauty, 238
  Carolina, 239
  Common, 240
Squaw-berry, 194
Squirrel-corn, 72
Starflower, 241
Star-grass, Yellow, 294
Starwort, 9
Strawbell, 184
Streptopus
  amplexifolius, 262, 263
  roseus, 262, 263
Summersweet, 27
Sundew, 207
Sweet William, Wild, 200, 201
Symplocarpus foetidus, 232

Taylor, Kathryn, 153
Teaberry, 290
Thalictrum
  dasycarpum, 179, 181
  dioicum, 179, 181
  polygamum, 179, 181
  venulosum, 179, 181
Thimbleweed, 7
Thoroughwort, Hyssop, 75, 76
Tiarella cordifolia, 94
Toad-shade, 252
Toothwort, 67
  Two-leaved, 67
Trailing-arbutus, viii, **91**
Trientalis borealis, 241
Trillium, 249, 252
  Dwarf White, 252
  Nodding, 251
  Painted, 252
  Purple, 251
  Red, 250, 251
  Rose, 250, 252
  Snow, 252
  Toad, 252
  White, 251
  Yellow, 252
Trillium
  cernuum, 249, 251
  erectum, 249, 251
  grandiflorum, 249, 251
  luteum, 249, 252
  nivale, 249, 252
  sessile, 249, 252
  stylosum, 249, 252
  undulatum, 249, 252
Trout-lily, 253–255
  Common, 255
  Eastern, 255
  Henderson's, 255
Trumpet-leaf, 208
Turtle-head, 256
  Pink, 257
  White, 257
Twin-berry, 194
Twin-flower, 258
Twin-leaf, 260
Twisted-stalk, 262

Unicorn-root, 82
Uvularia
  grandiflora, 182, 183
  perfoliata, 182, 184
  sessilifolia, 182, 184

Venus Fly-trap, 207
Verbena, Large-flowered, 268
Verbena
  canadensis, 267, 269
  hastata, 267, 269
  stricta, 267, 269
Vernonia
  altissima, 144
  noveboracensis, 144
Veronicastrum virginicum, 68
Vervain, 267–269
  Blue, 268, 269
  Erect, 268, 269
Viola
  adunca, 270, 272

blanda, 270, 273
canadensis, 270, 273
conspersa, 270, 273
eriocarpa, 270, 273
palmata, 270, 273
pedata, 270, 273
pedatifida, 270, 274
priceana, 270, 274
pubescens, 270, 274
rotundifolia, 270, 274
Violet, 270–274
American Dog, 273
Bird's-foot, 273
Canada, 273
Confederate, 274
Downy Yellow, 274
Early Blue, 273
Prairie, 274
Round-leaved Yellow, 274
Sand, 272
Smooth Yellow, 273
Sweet White, 273
Virginia-bluebells, 275
Virginia-cowslip, 276

Wake-robin, 250, 251
Water-arum, 279
White-hearts, 72
Wild-oats, 184
Wild-ginger, 282
Mottled, 284
Wild-petunia, 285
Wild Senna, 197
Wild Sweet William, 200, 201
Windflower, 4, 6
American Wood, 6
Canada, 6
Carolina, 6
Wine-poppy, 287
Wing-stem, 34
Wine-cups, 288
Wintergreen, 65, 289
Spotted, 205
Wood-betony, 292
Wood-lily, Speckled, 17

Yellow Star-grass, 294